BIO-ART™ TO ACCOMPANY

Biology

THIRD EDITION

SOLOMON • BERG • MARTIN • VILLEE

SAUNDERS COLLEGE PUBLISHING

Harcourt Brace College Publishers

Fort Worth • Philadelphia • Boston • New York • Chicago • Orlando
San Francisco • Atlanta • Dallas • London • Toronto • Austin • San Antonio

Printed in the United States of America.

Solomon/Berg/Martin/Villee: Bio-Art to accompany BIOLOGY, 3/E by Solomon, Berg, Martin, and Villee

ISBN 0-03-096844-5

345 021 98765432

A Note to the Instructor

Thank you for adopting *Biology,* Third Edition by Solomon, Berg, Martin, and Villee. We appreciate your use of our text and as a special aid for students we have produced Bio-Art™.

Bio-Art™ is a collection of important pieces of art from *Biology,* Third Edition rendered in black and white. Generally most pieces of Bio-Art™ do not include labels, so students can use the art as a learning tool. It is an excellent tool in measuring student understanding of processes and organisms.

This valuable study aid is free with copyright privileges to instructors who adopt of *Biology,* Third Edition for class or can be purchased by students at a low cost. Please contact your bookstore if you would like Bio-Art™ to be purchased by students as a required or recommended supplement.

Suggested uses of Bio-Art™, for instructors who adopt *Biology,* Third Edition, include:

- ✔ Copying Bio-Art™ as handouts for students, enabling students to label parts of figures, take notes, and avoid redrawing complicated diagrams, while the instructor uses a corresponding color overhead transparency or slide from *Biology,* Third Edition

- ✔ Duplicating Bio-Art™ on overhead acetates and referring to and writing on the acetates during lectures, while students refer to their own handouts

- ✔ Distributing copies of Bio-Art™ as part of exams and quizzes

- ✔ Having students complete homework assignments using their own copies of Bio-Art™

BIO-ART

Biology, 3/e
Solomon/Berg/Martin/Villee

Figure #	Description
4-8	Comparison of light and electron microscopes
4-13	Plant cell
4-14	Animal cell
4-17 a-c	The cell nucleus and the nuclear envelope
4-20	The endomembrane system
4-31 a,b,d	Structure of a cilium
4-34	Elements of the cytoskeleton
5-2	Membrane structure
5-9	Transport pathway
5-10	Intrinsic membrane proteins
5-17a	Sodium-potassium ATPase
5-17b	Sodium-potassium pump
5-18	Coupled glucose transport
5-20 a-d	Phagocytosis
5-22	Receptor-mediated endocytosis
5-23	Desmosome
5-25	Tight junction
5-26 a-d	Gap junction
7-4	Respiration phases
7-13	Catabolism
7-15	Fermentation
8-2b	Chloroplast envelope
8-5	Light reactions

**Light
microscope**

**Transmission
electron
microscope**

**Scanning
electron
microscope**

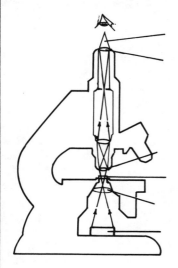
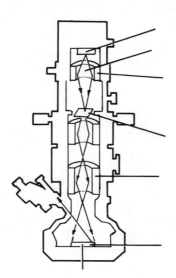
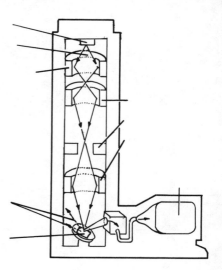

NOTES

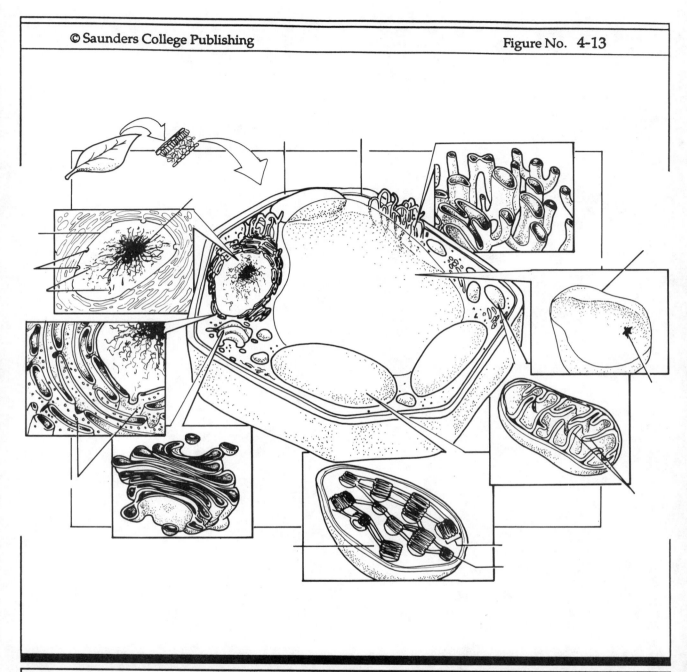

NOTES

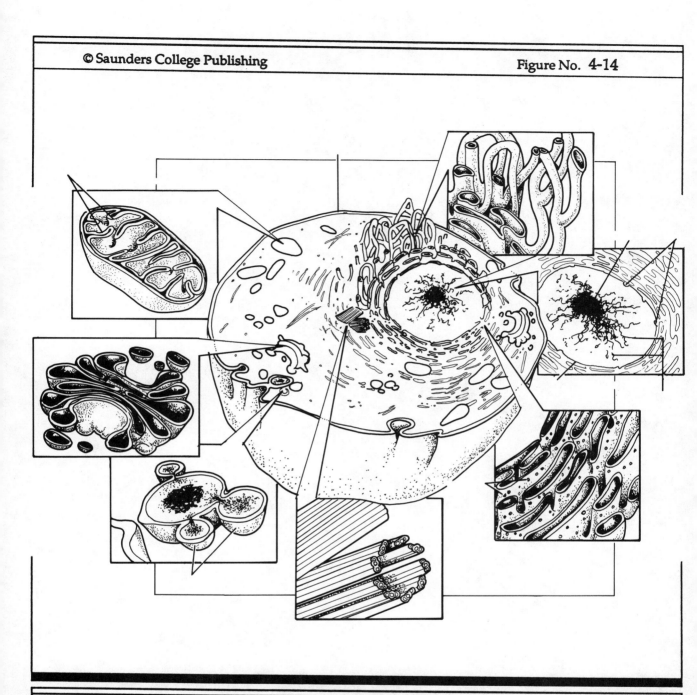

NOTES

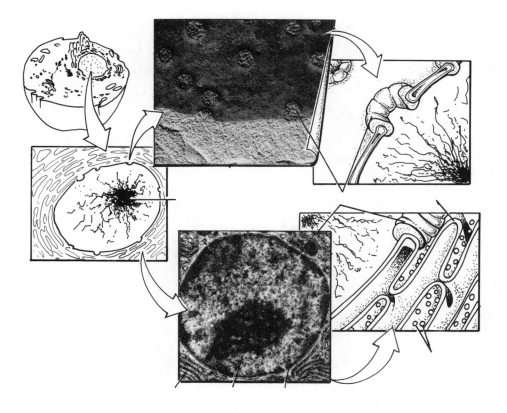

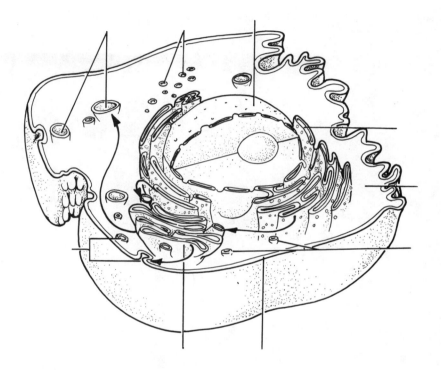

NOTES

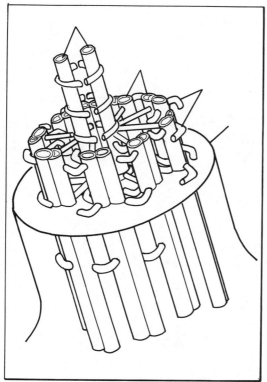

a

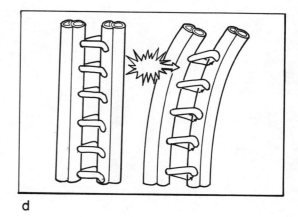

d

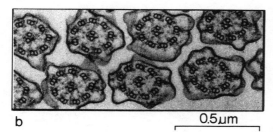

b   0.5μm

Figure No. 4-34

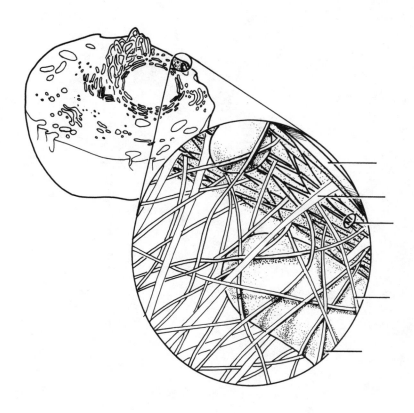

NOTES

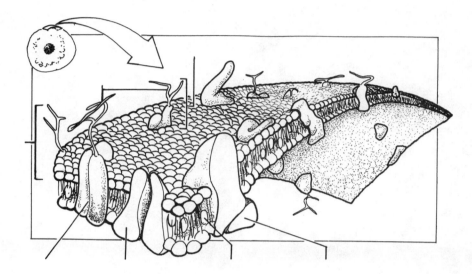

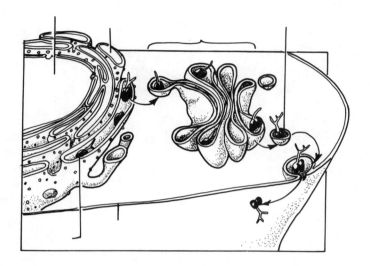

NOTES

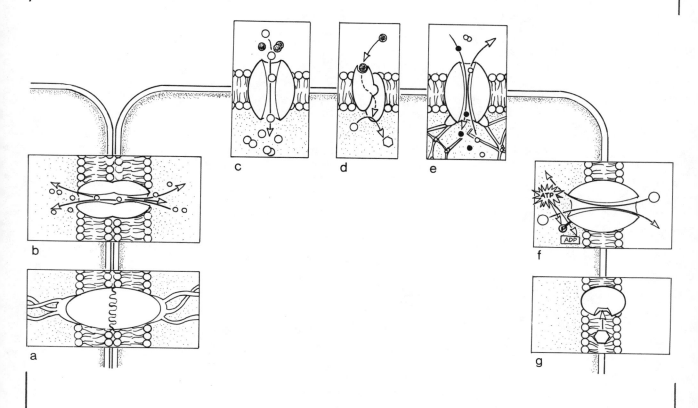

b

a

c

d

e

f

g

NOTES

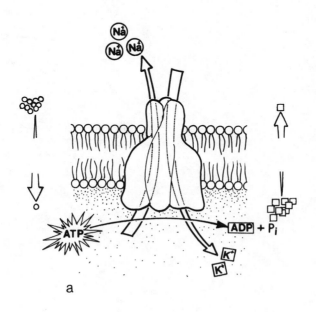

a

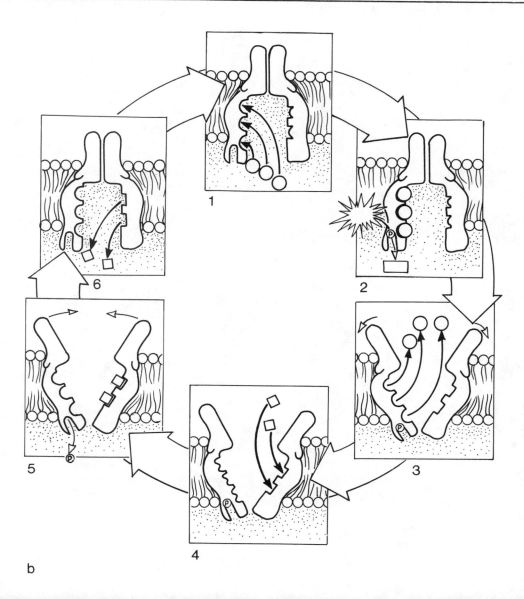

b

NOTES

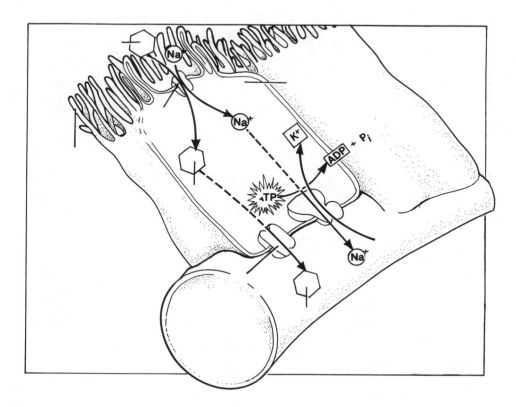

NOTES

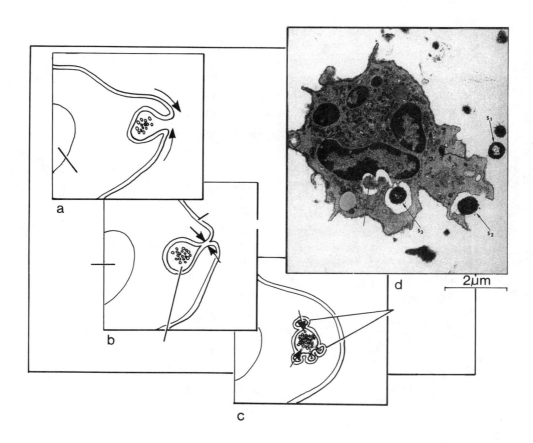

a

b

c

d

2µm

NOTES

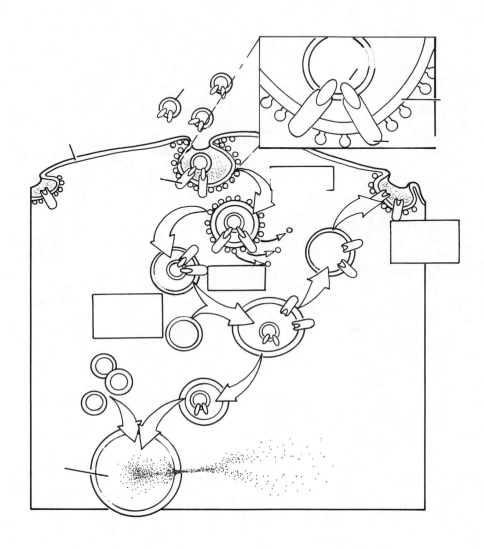

NOTES

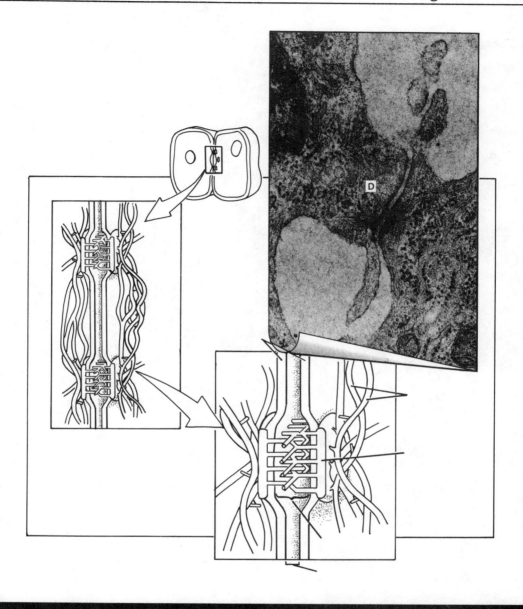

NOTES

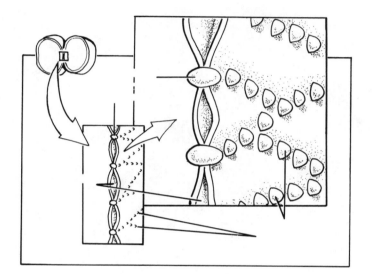

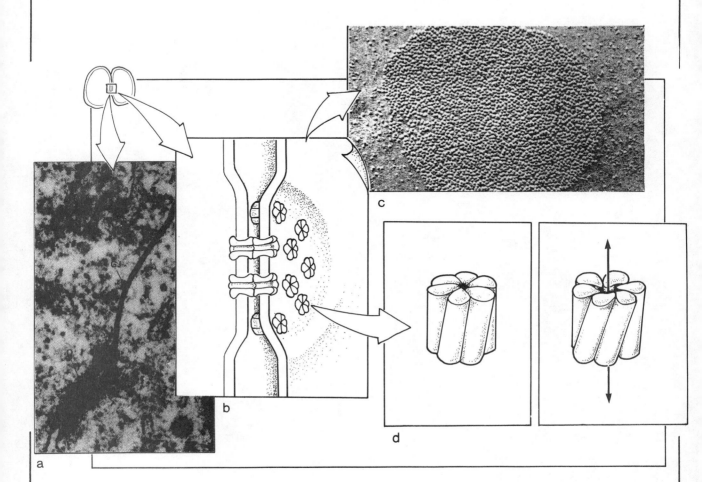

a

b

c

d

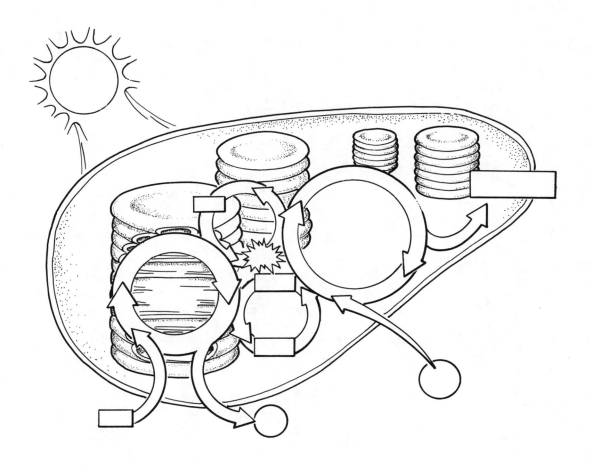

NOTES

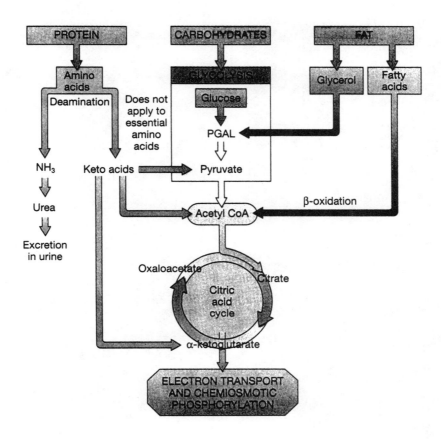

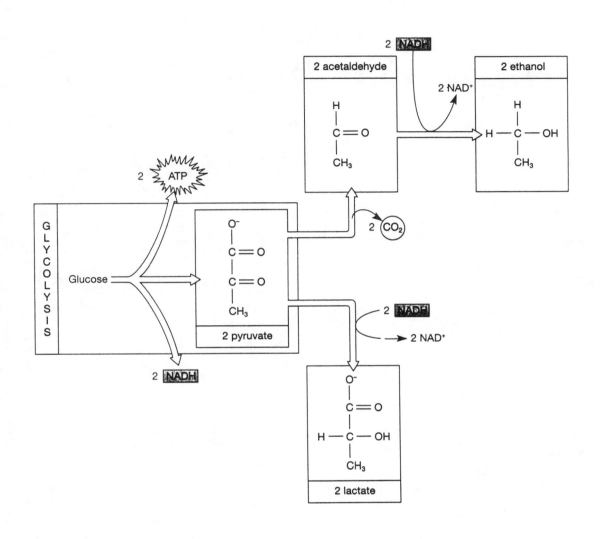

NOTES

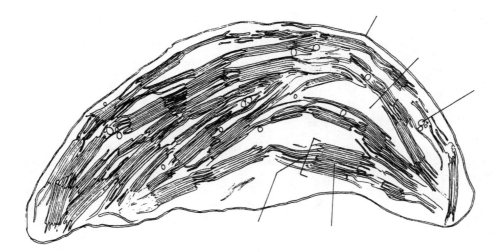

NOTES

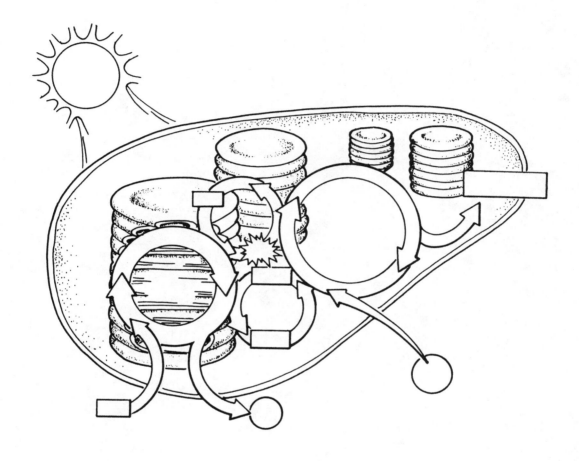

NOTES

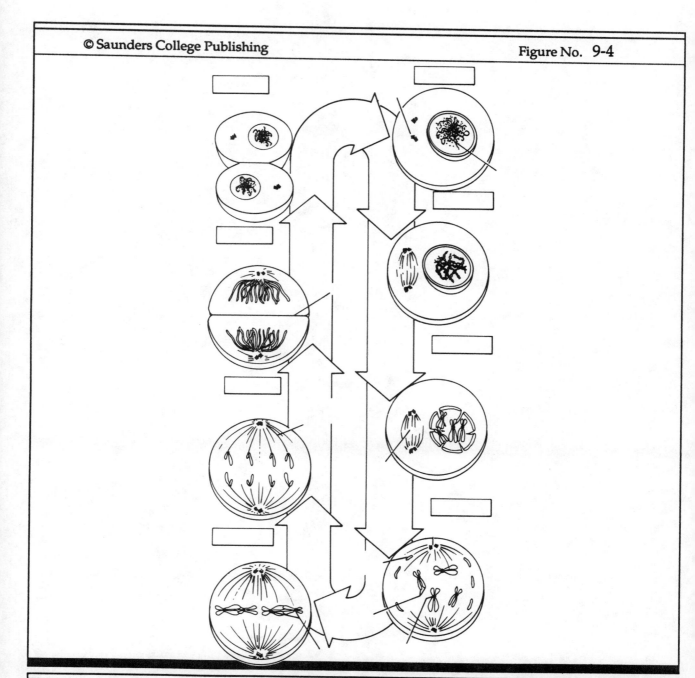

NOTES

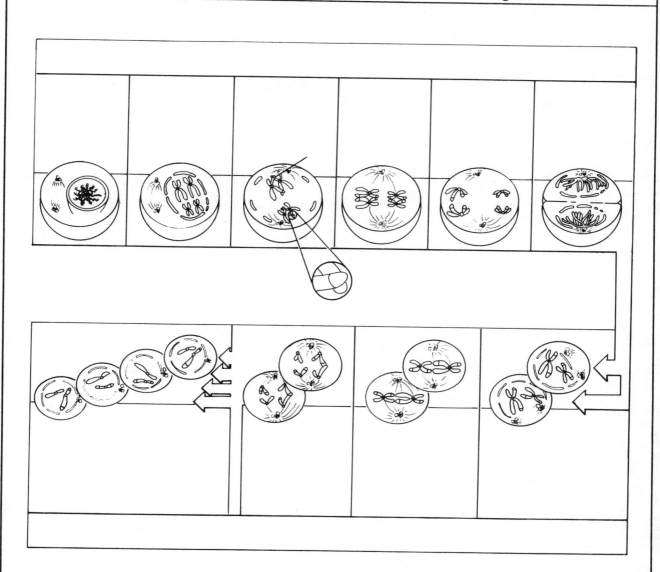

NOTES

NOTES

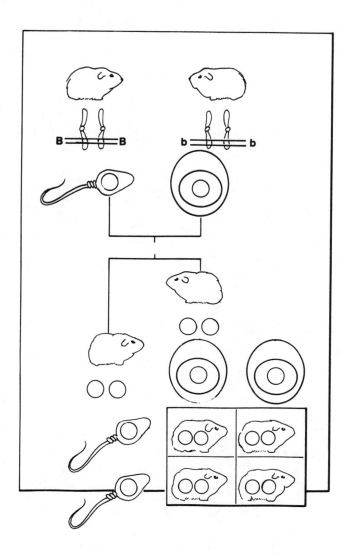

NOTES

Figure No. 10-5

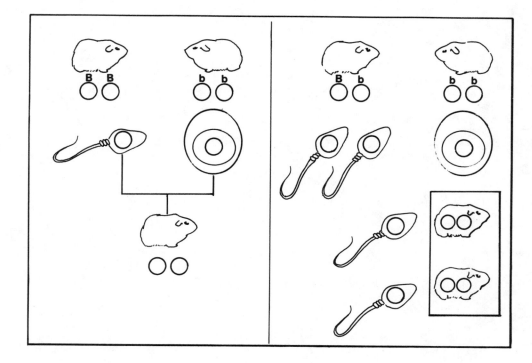

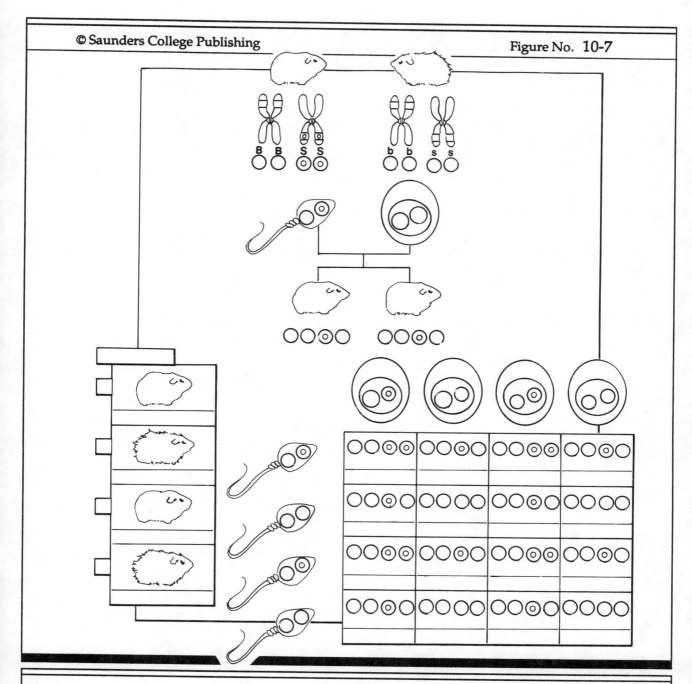

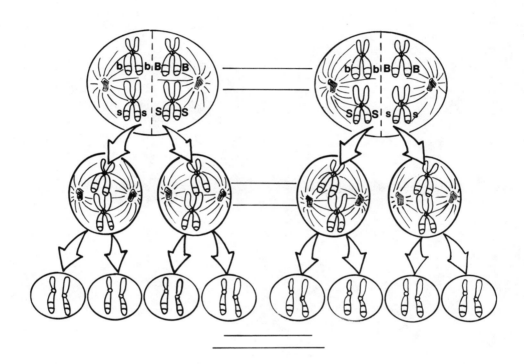

NOTES

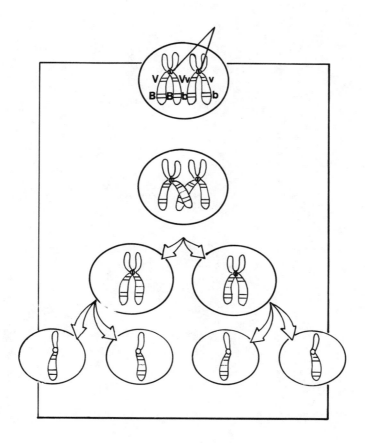

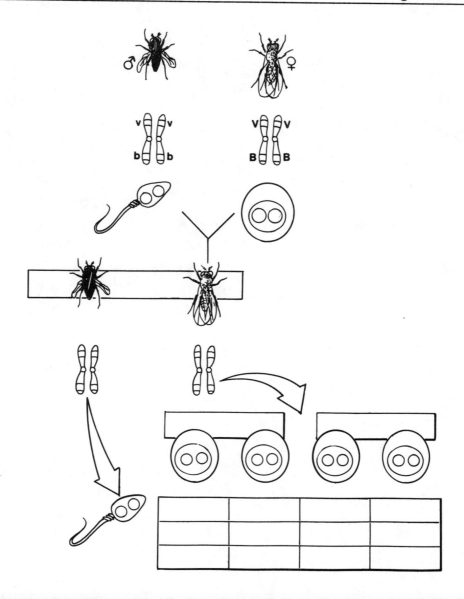

NOTES

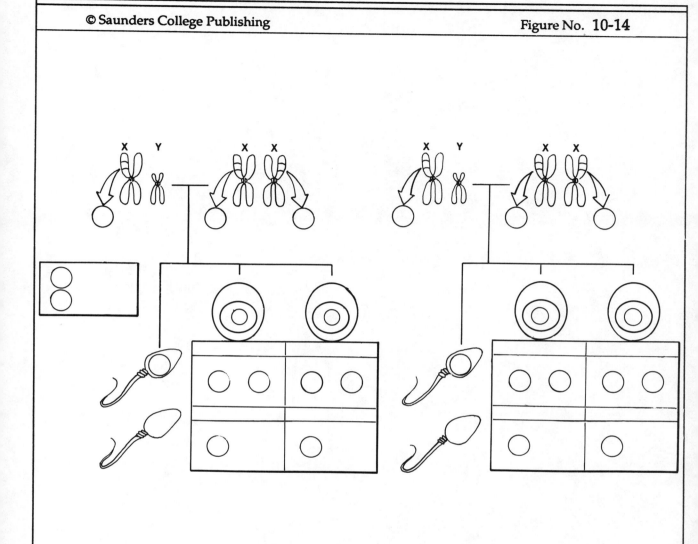

NOTES

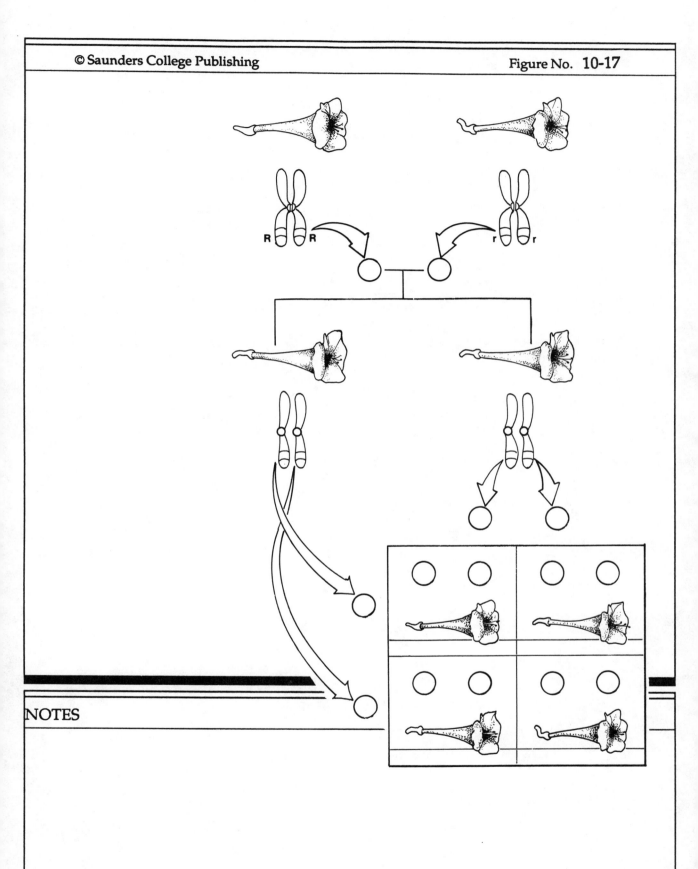

NOTES

Figure No. 10-20 a

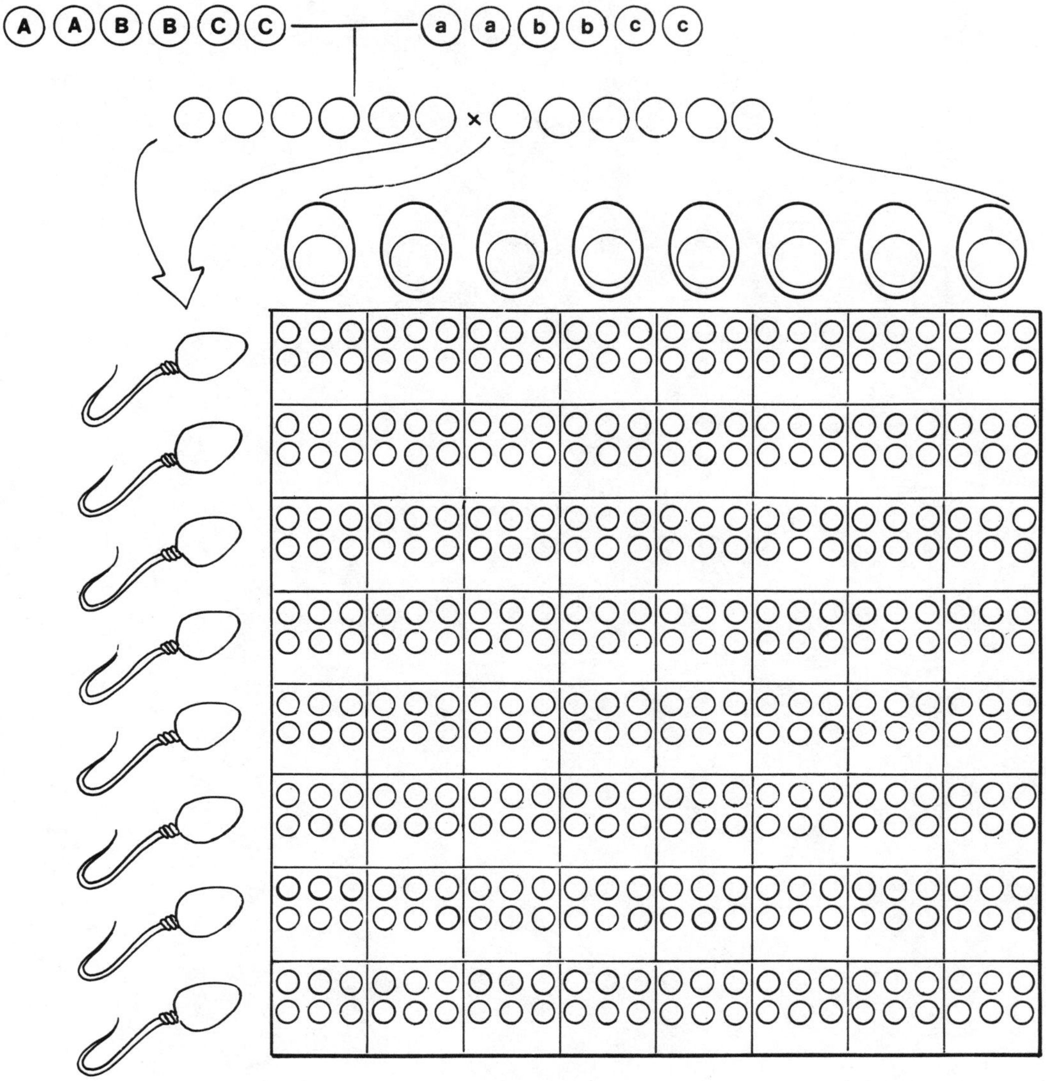

NOTES

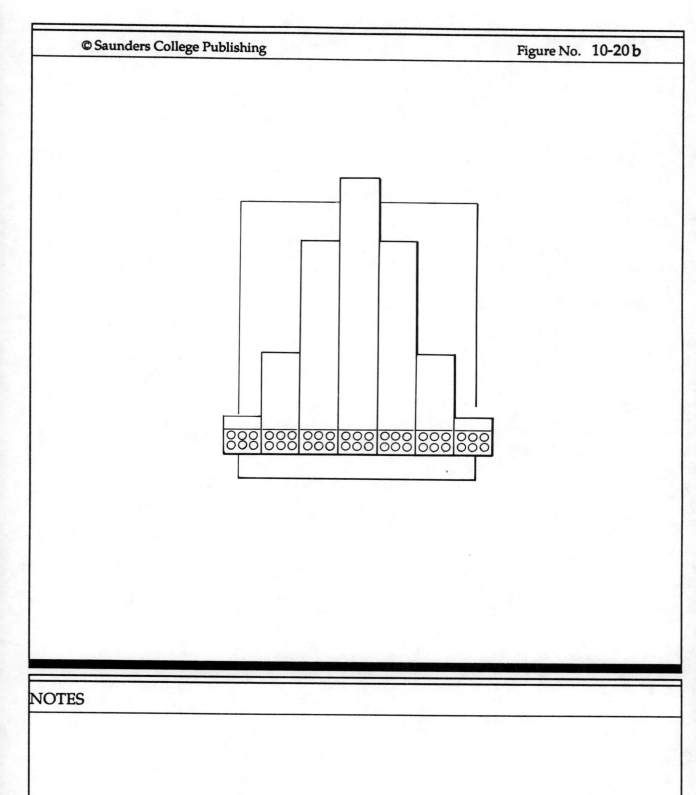

NOTES

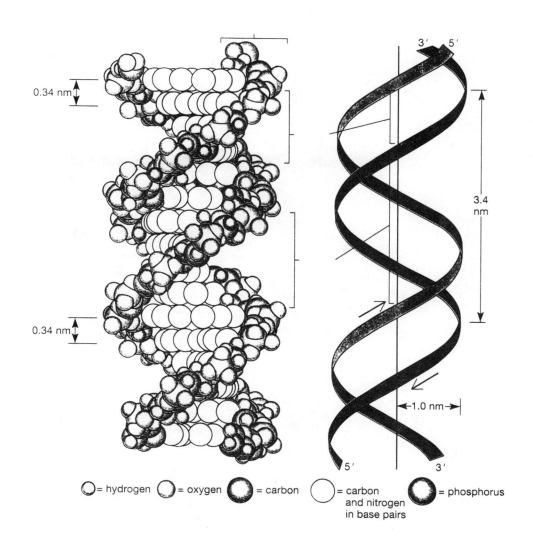

0.34 nm

0.34 nm

3' 5'

3.4 nm

1.0 nm

5' 3'

○ = hydrogen   ○ = oxygen   ● = carbon   ○ = carbon and nitrogen in base pairs   ● = phosphorus

NOTES

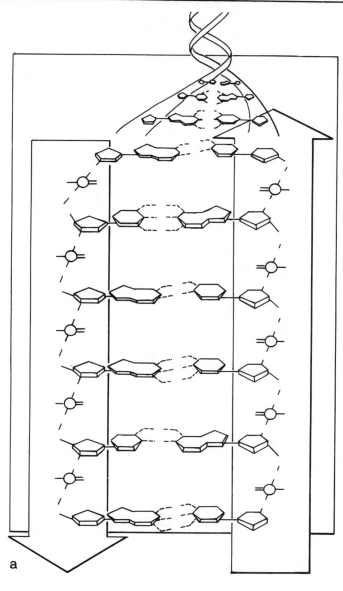

a

NOTES

b

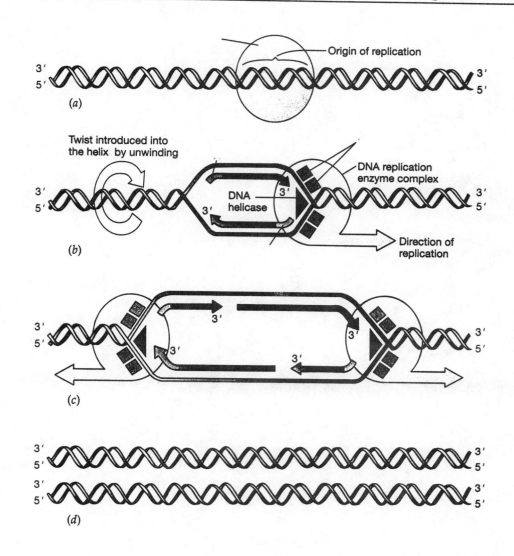

Origin of replication

(a)

Twist introduced into
the helix by unwinding

DNA replication
enzyme complex

DNA
helicase

Direction of
replication

(b)

(c)

(d)

NOTES

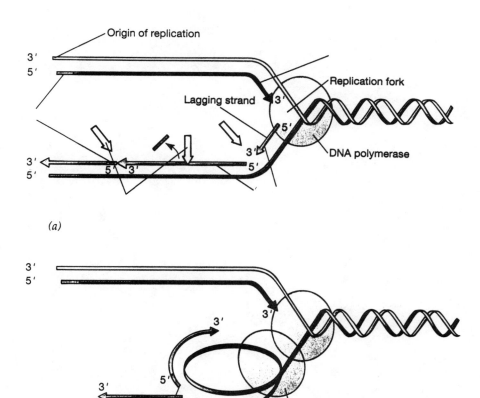

(a)

(b)

NOTES

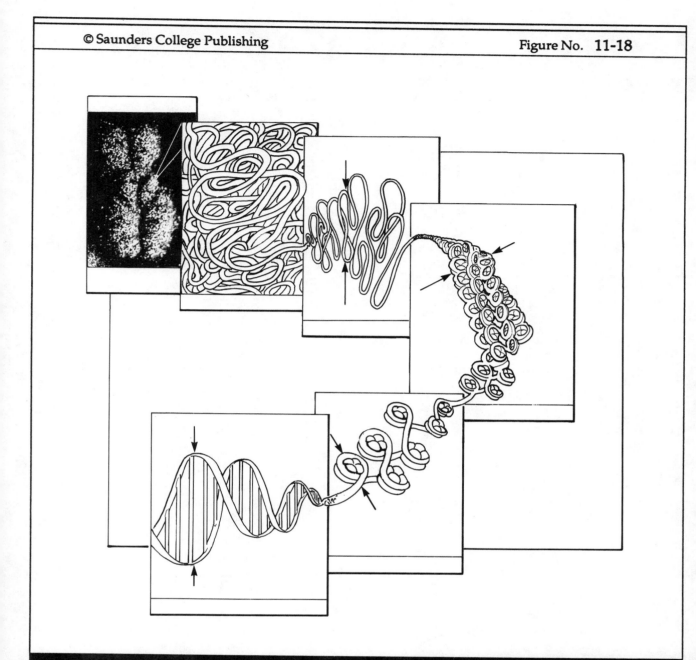

NOTES

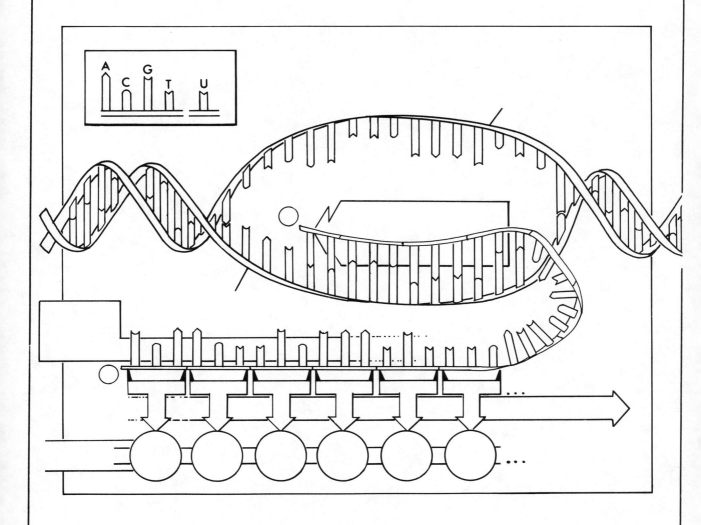

NOTES

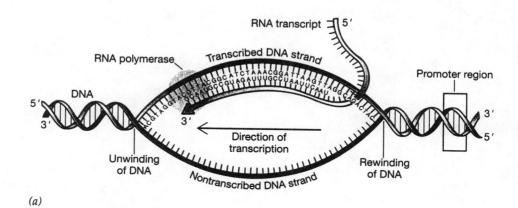

(a)

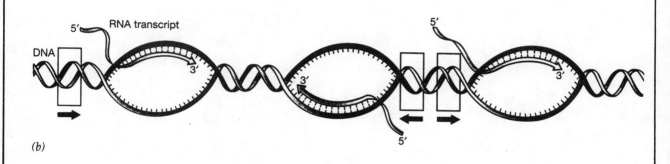

(b)

NOTES

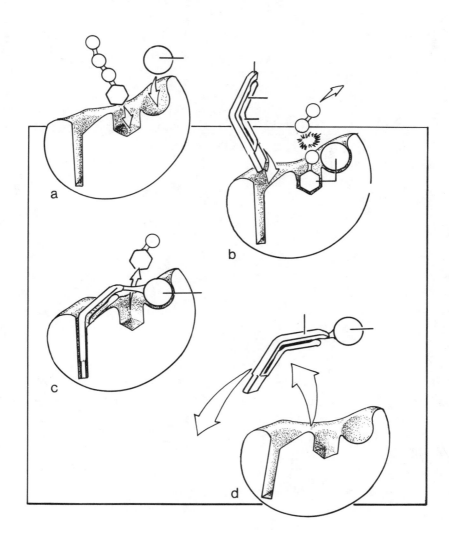

NOTES

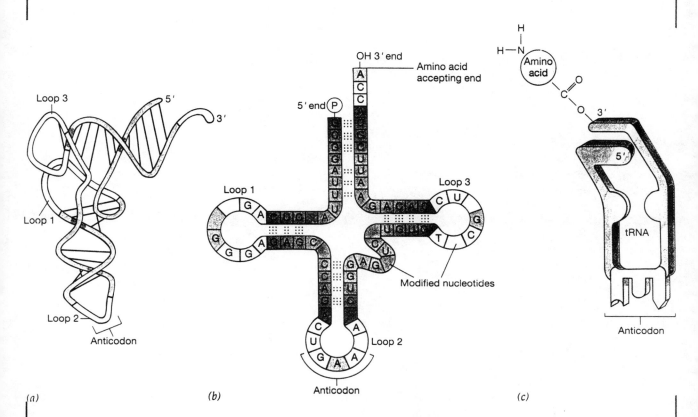

(a)

(b)

(c)

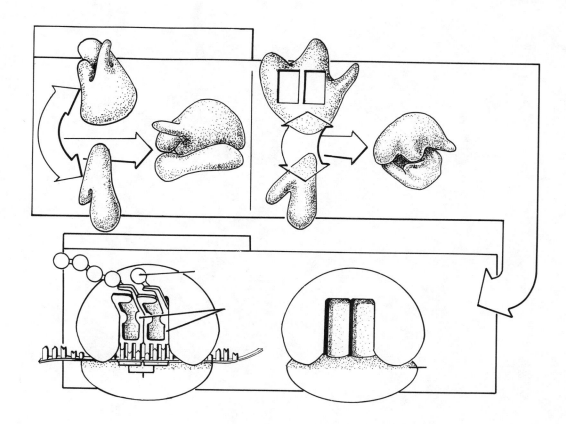

NOTES

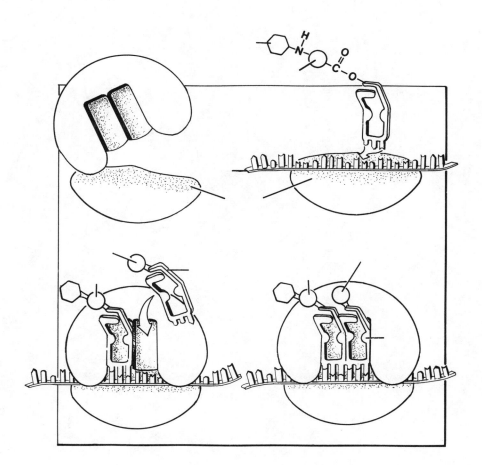

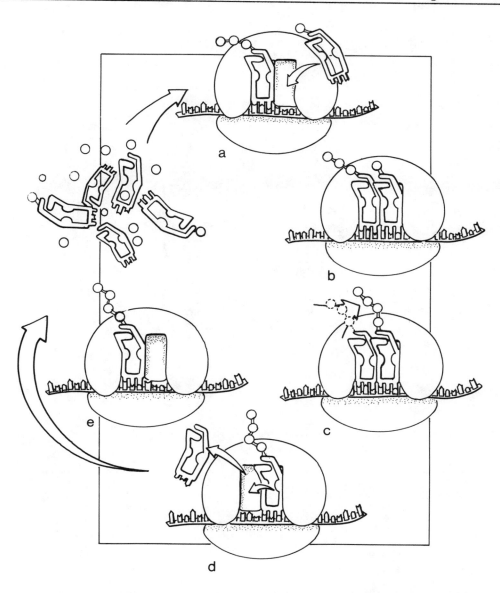

NOTES

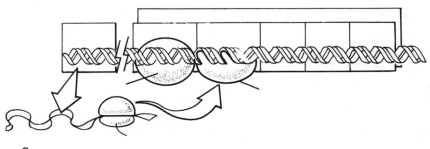

a

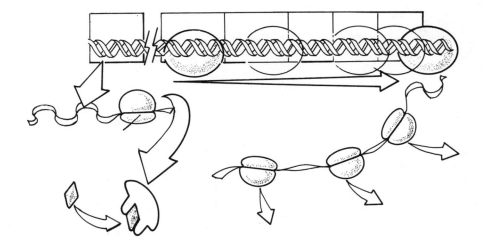

b

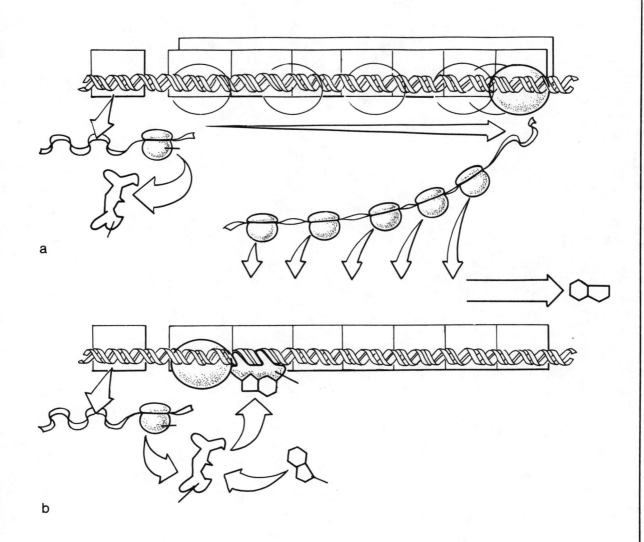

a

b

NOTES

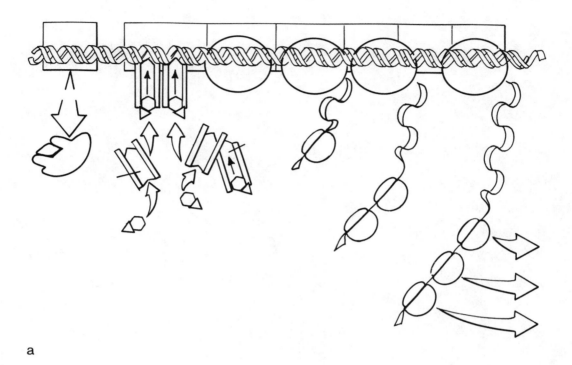

a

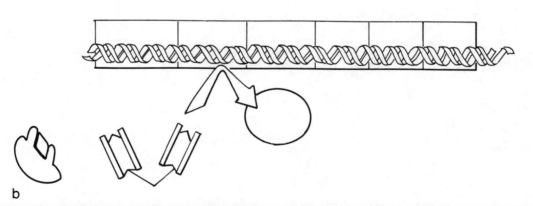

b

NOTES

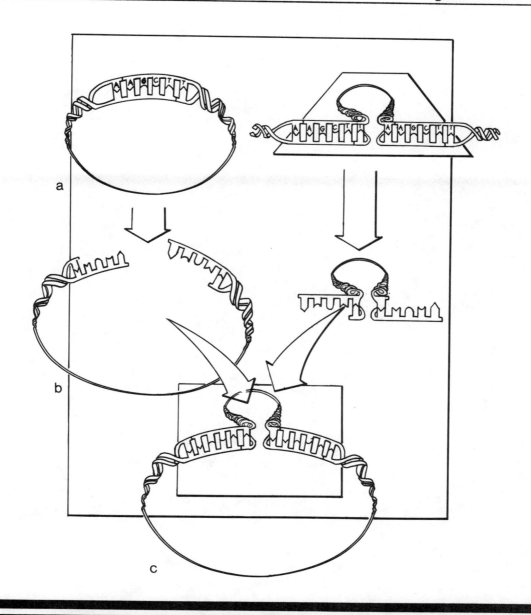

a

b

c

NOTES

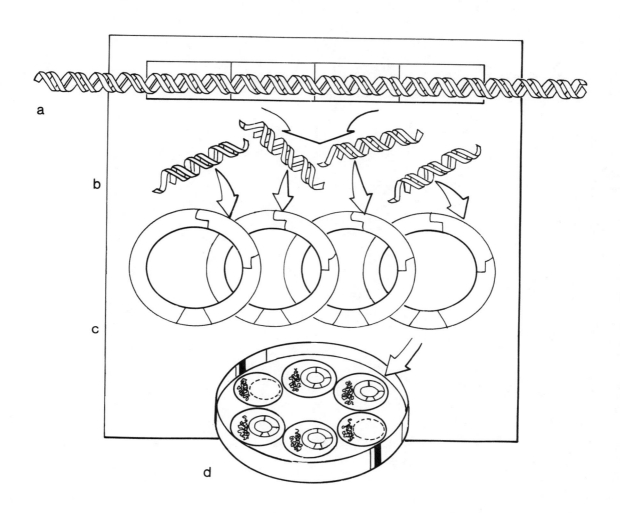

a

b

c

d

NOTES

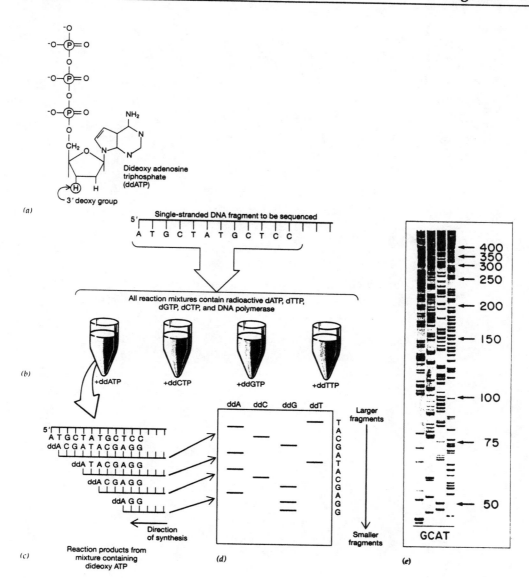

(a) Dideoxy adenosine triphosphate (ddATP)

3' deoxy group

Single-stranded DNA fragment to be sequenced

5'   A T G C T A T G C T C C

All reaction mixtures contain radioactive dATP, dTTP, dGTP, dCTP, and DNA polymerase

(b)   +ddATP   +ddCTP   +ddGTP   +ddTTP

(c)
5'
A T G C T A T G C T C C
ddA C G A T A C G A G G
ddA T A C G A G G
ddA C G A G G
ddA G G

← Direction of synthesis

Reaction products from mixture containing dideoxy ATP

(d)
ddA   ddC   ddG   ddT

Larger fragments

T A C G A T A C G A G G

Smaller fragments

(e)

GCAT

← 400
← 350
← 300
← 250
← 200
← 150
← 100
← 75
← 50

NOTES

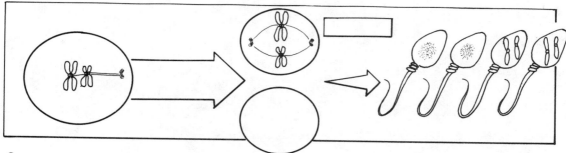

a

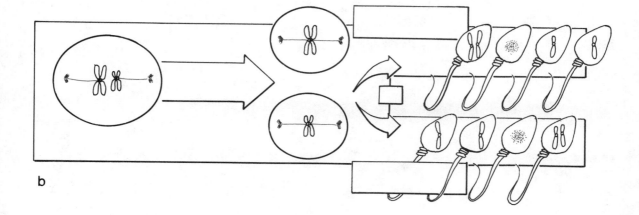

b

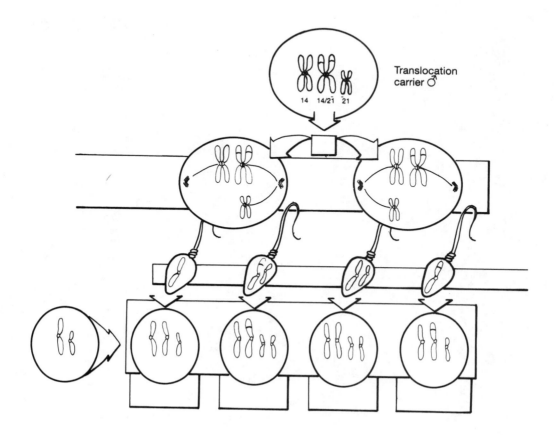

Translocation carrier ♂

NOTES

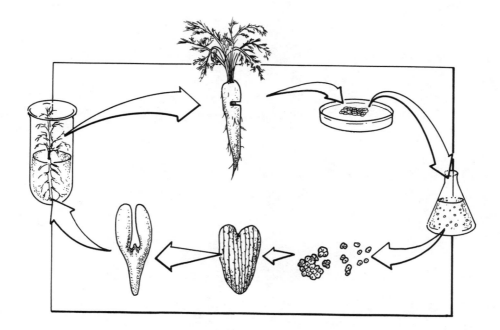

NOTES

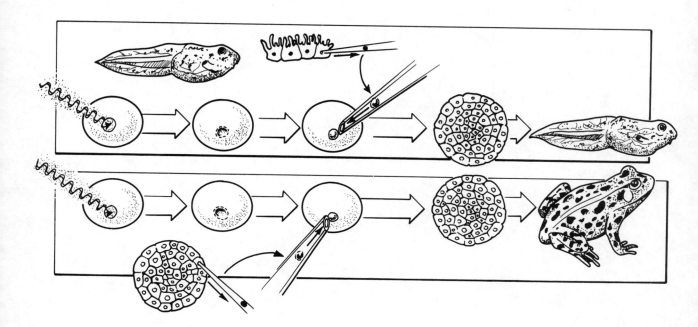

NOTES

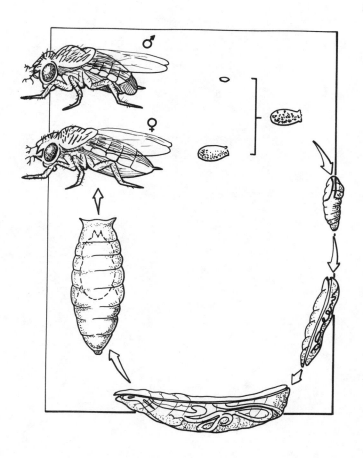

NOTES

Figure No. **16-7**

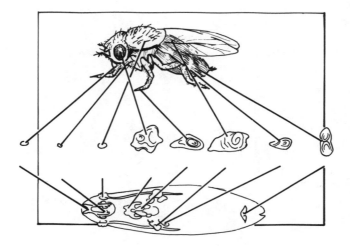

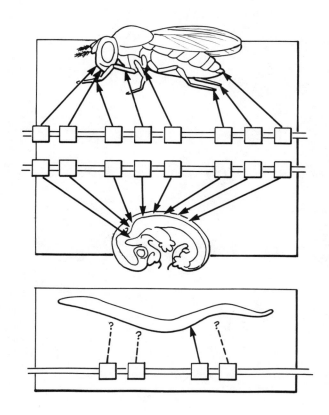

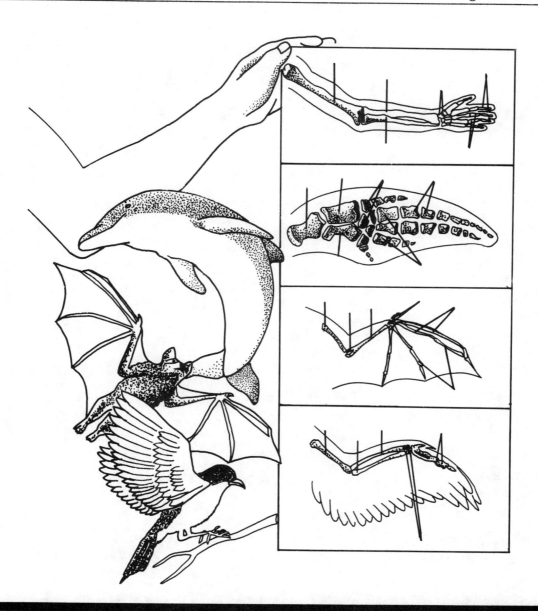

NOTES

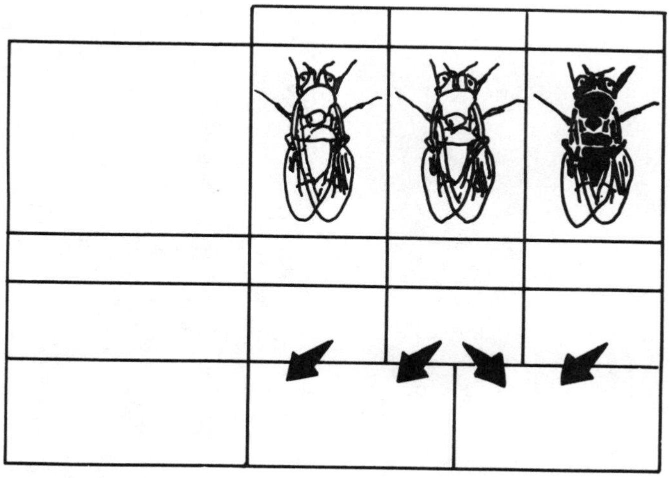

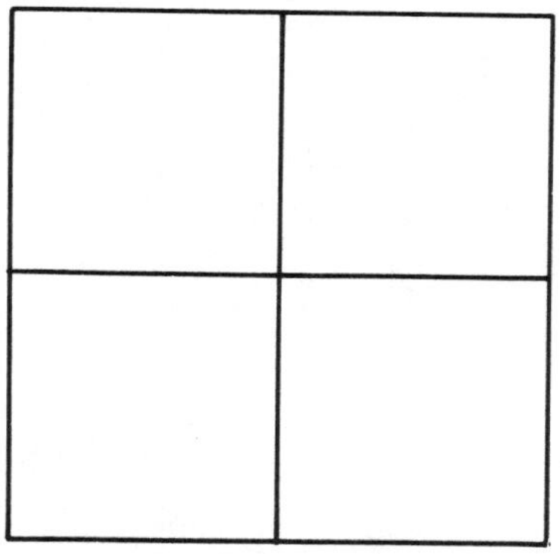

NOTES

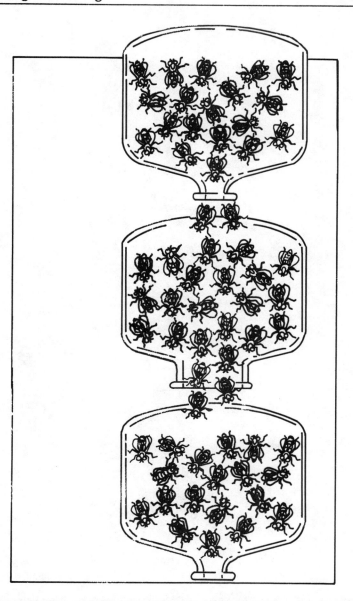

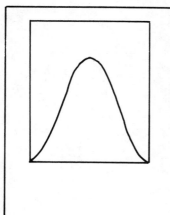

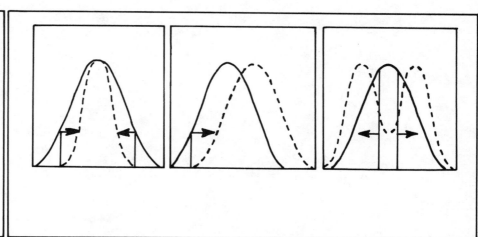

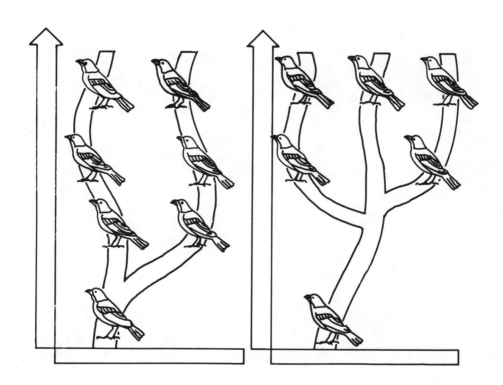

NOTES

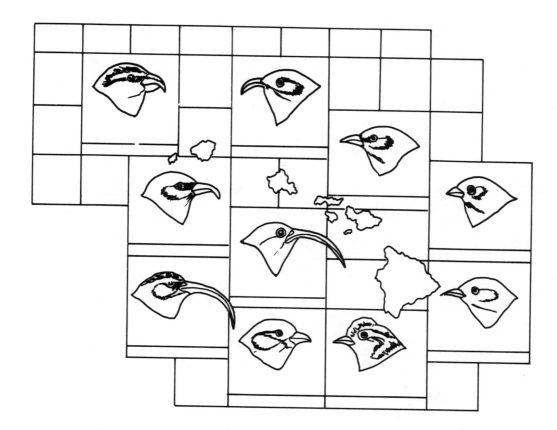

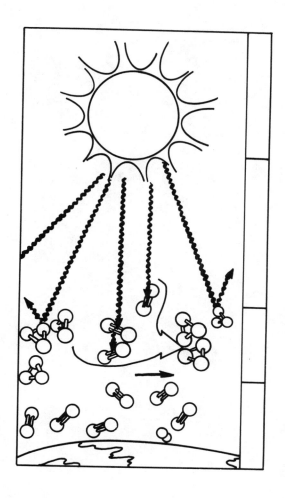

NOTES

NOTES

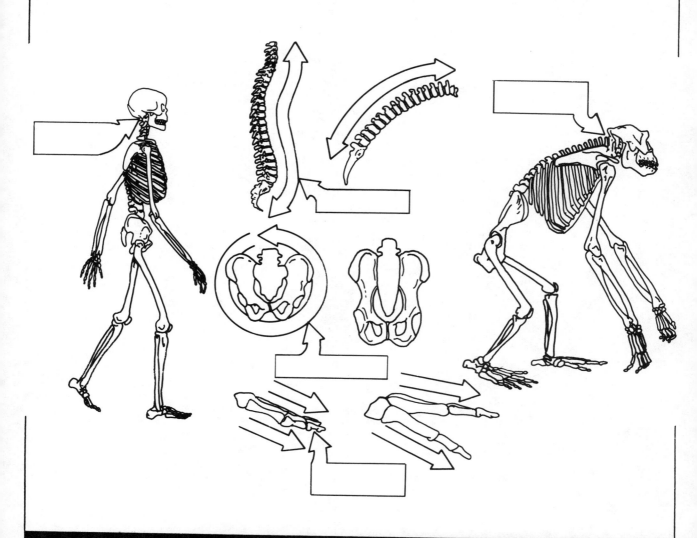

NOTES

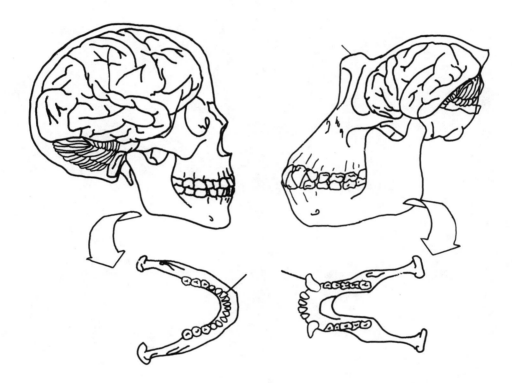

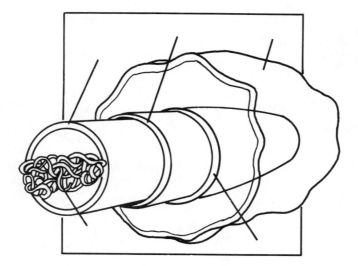

NOTES

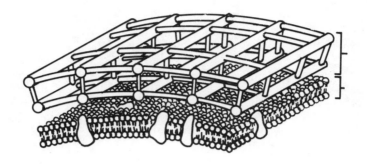

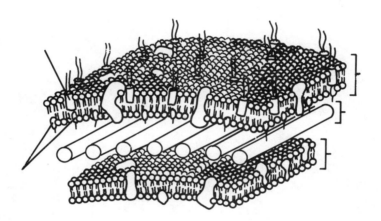

NOTES

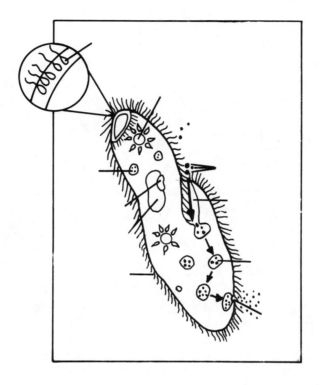

NOTES

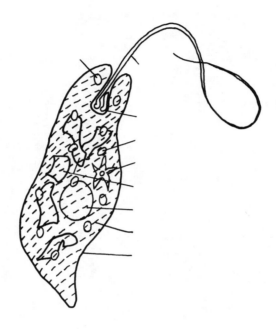

NOTES

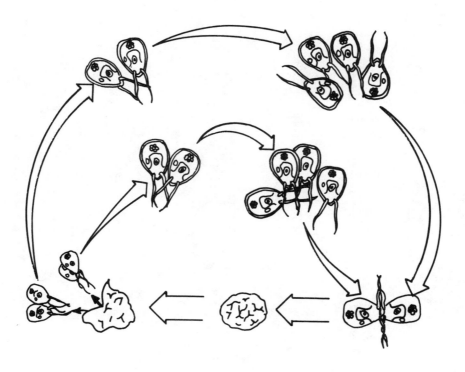

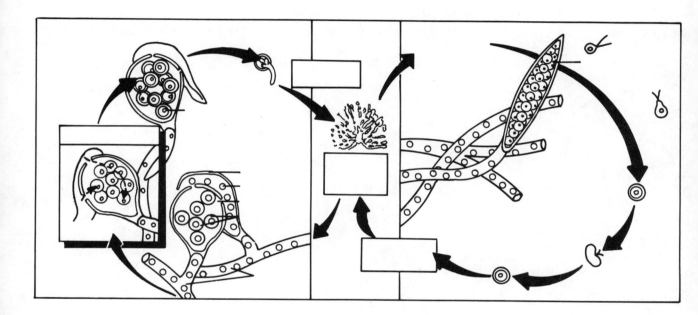

NOTES

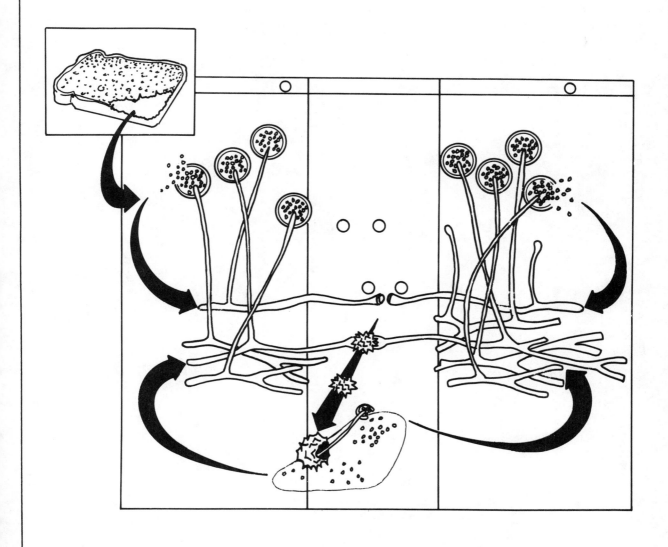

NOTES

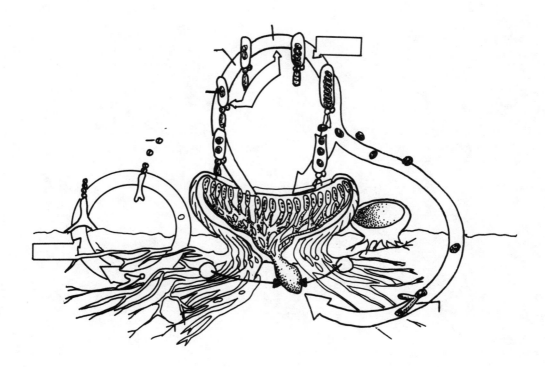

NOTES

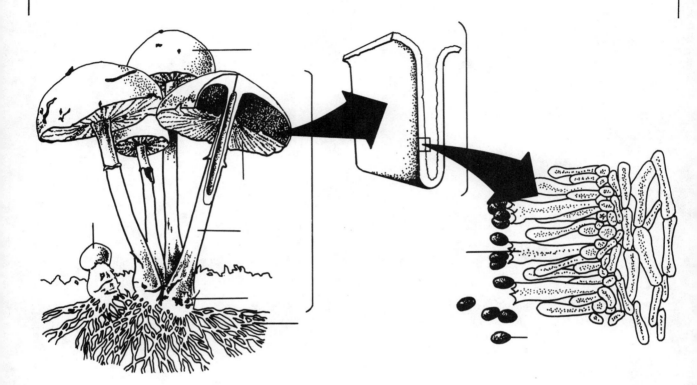

NOTES

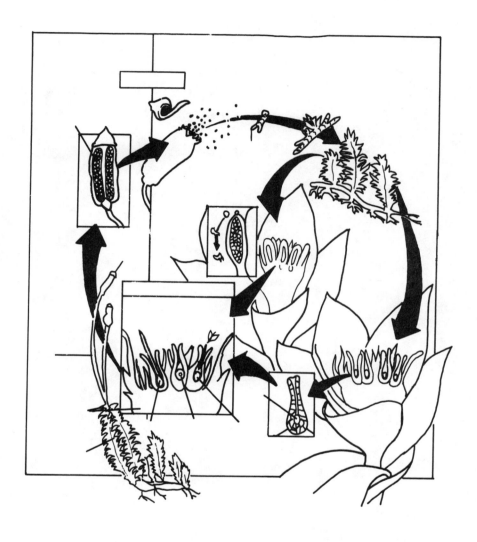

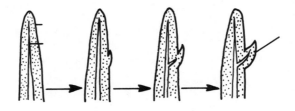

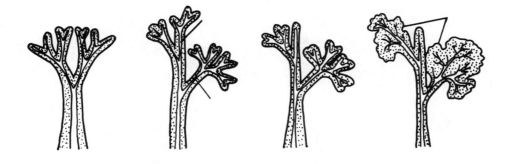

NOTES

NOTES

NOTES

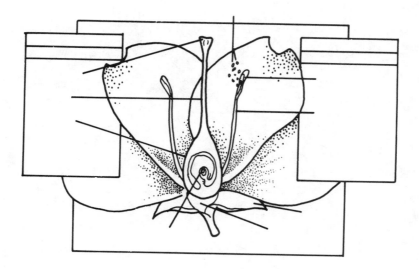

NOTES

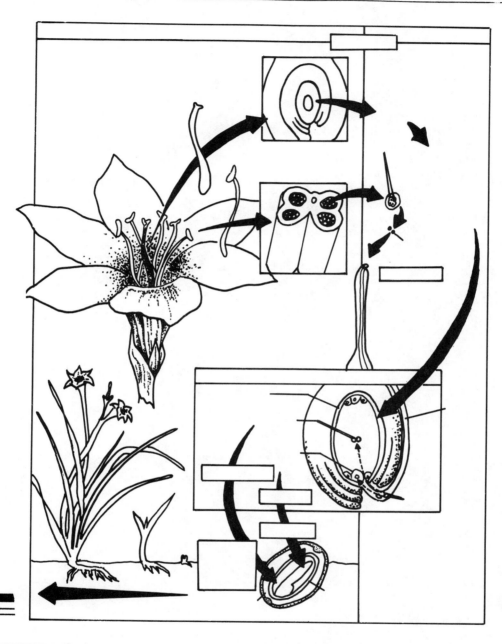

NOTES

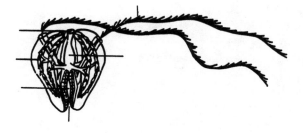

NOTES

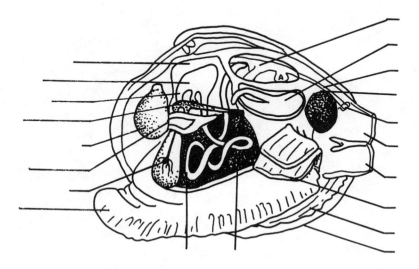

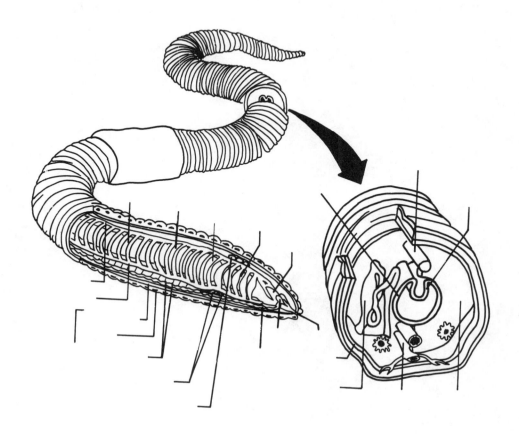

NOTES

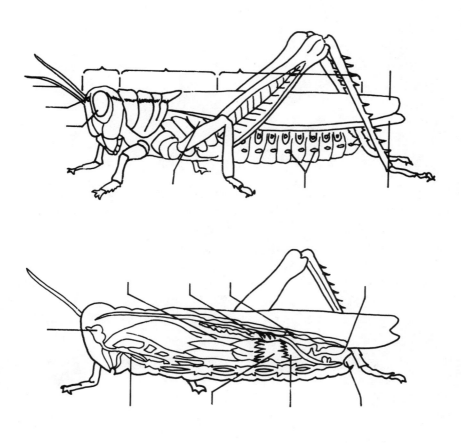

NOTES

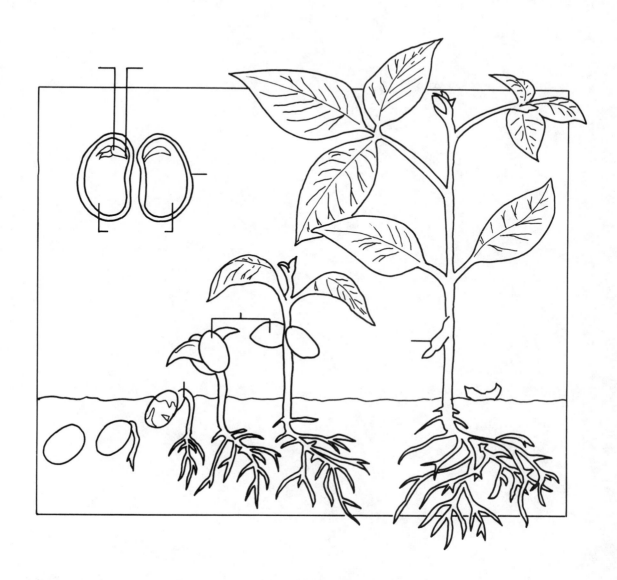

NOTES

Figure No. **31-3**

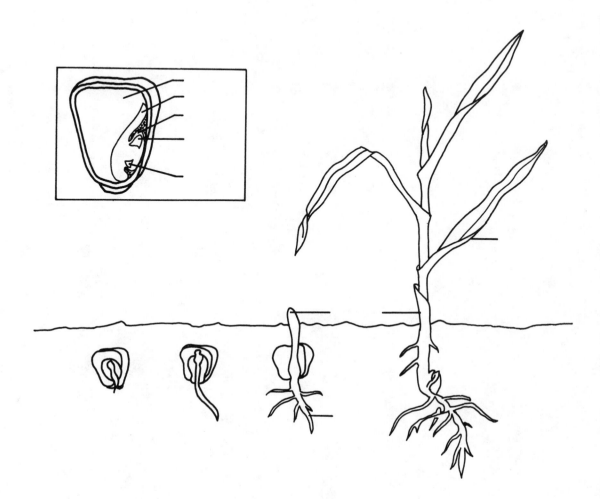

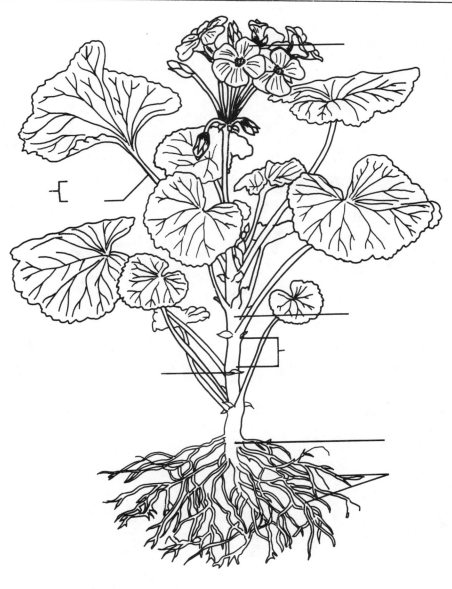

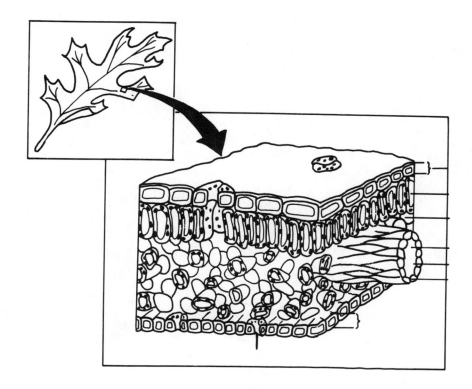

NOTES

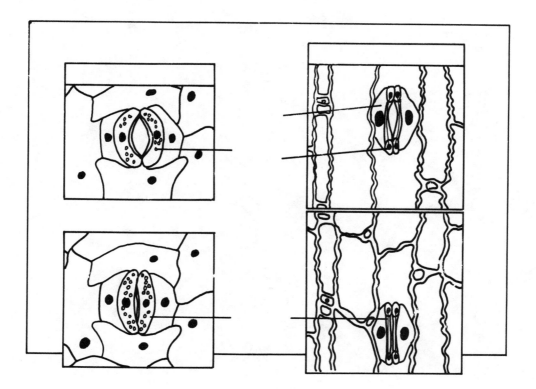

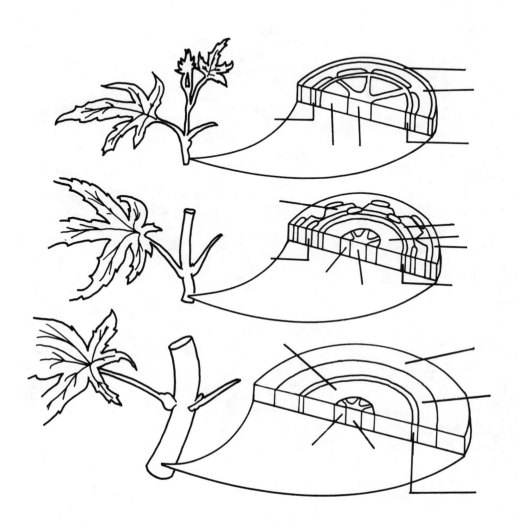

NOTES

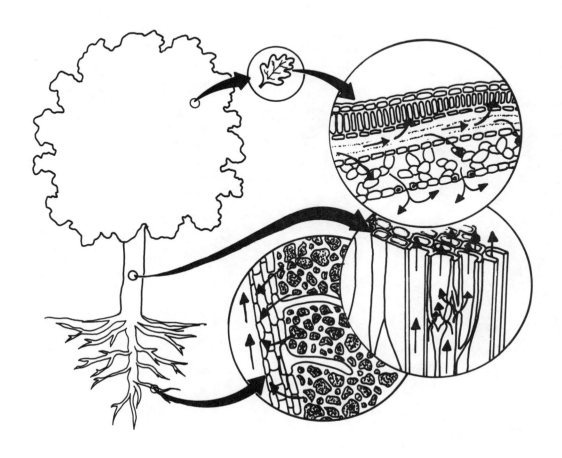

NOTES

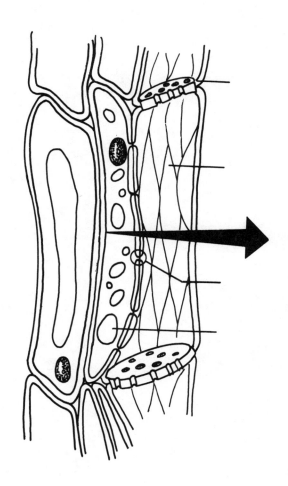

NOTES

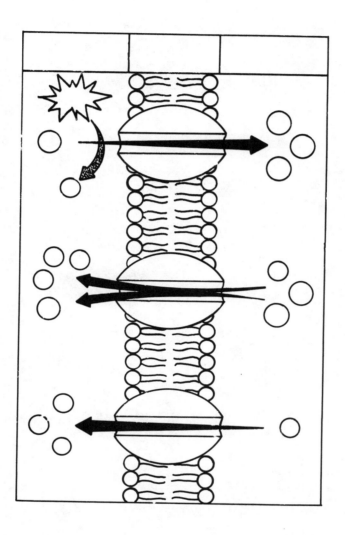

NOTES

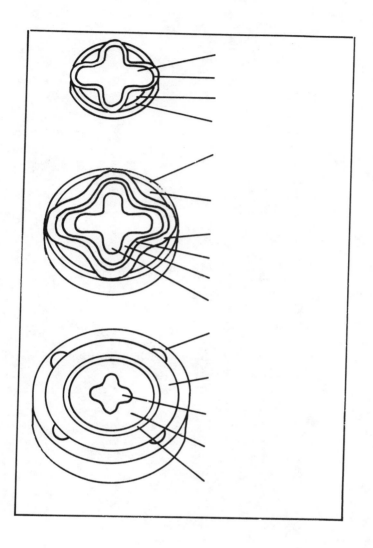

NOTES

Figure No. **35-8 b, f**

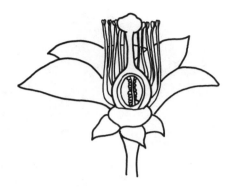

NOTES

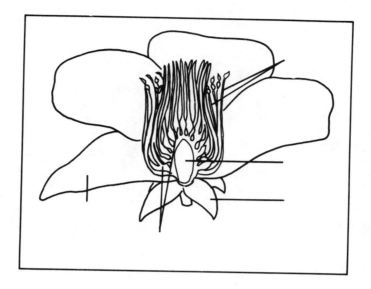

NOTES

Figure No. 35-14 a, b

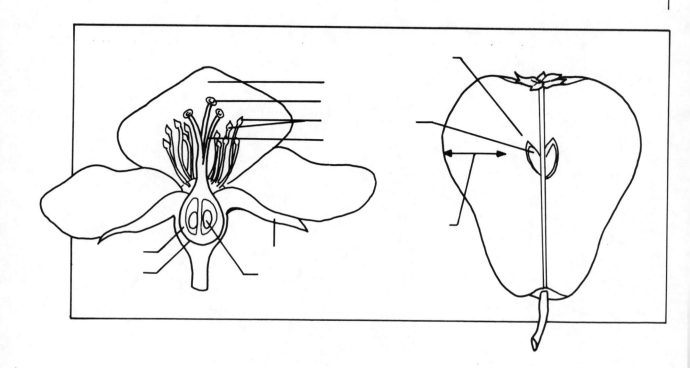

NOTES

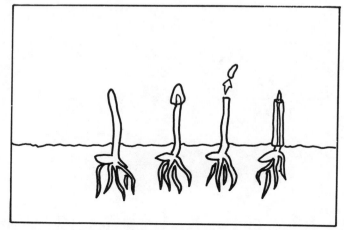

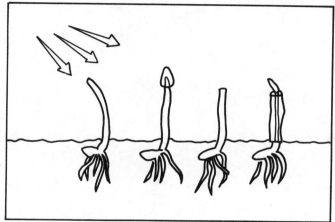

NOTES

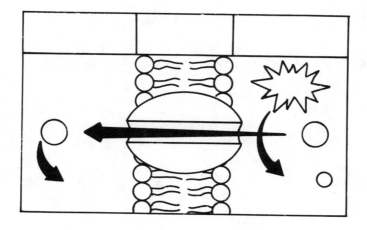

NOTES

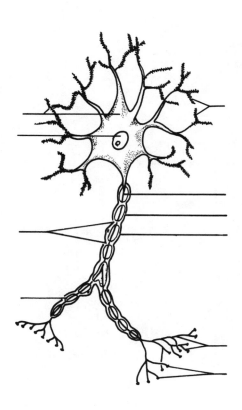

NOTES

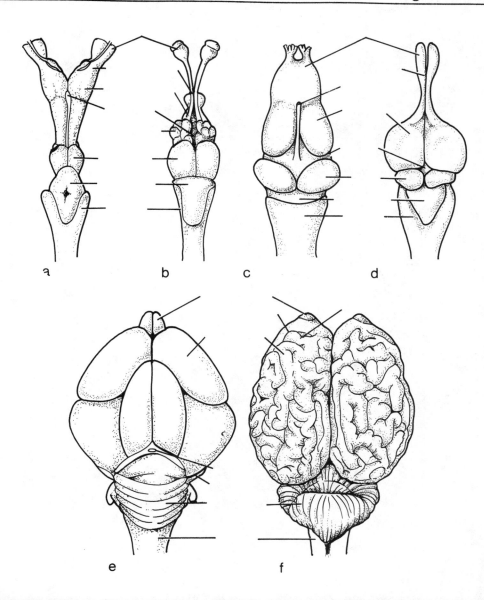

a

b

c

d

e

f

NOTES

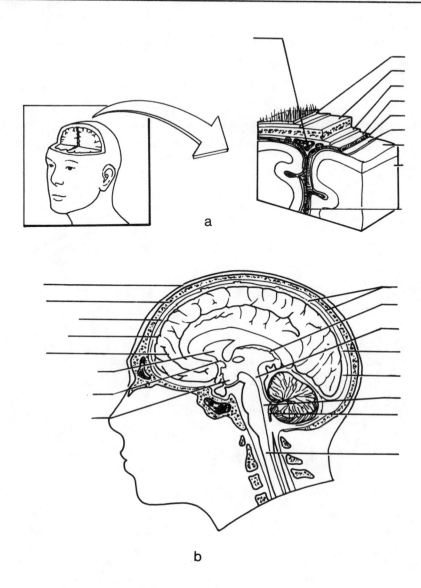

a

b

NOTES

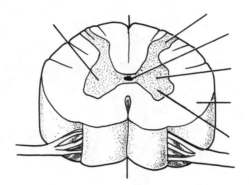

NOTES

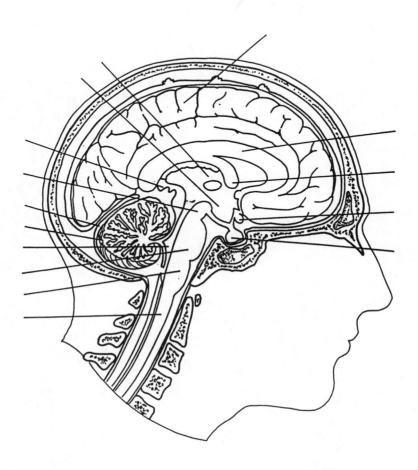

NOTES

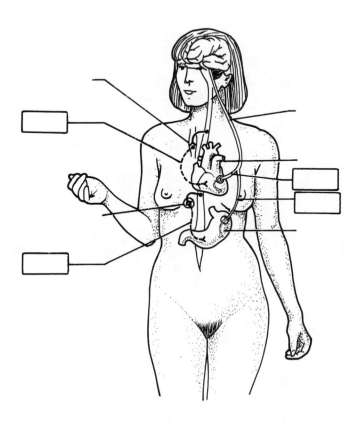

NOTES

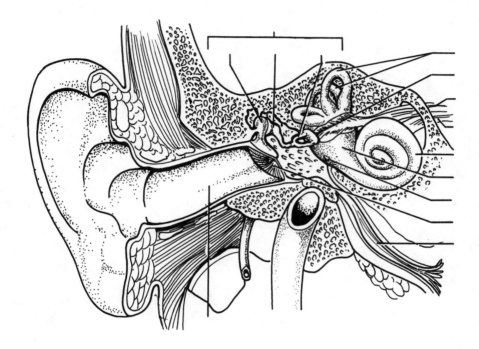

NOTES

NOTES

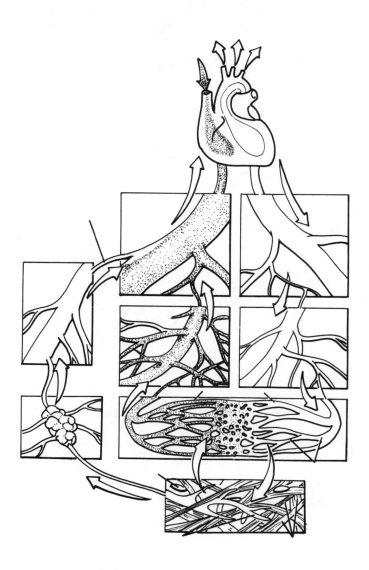

NOTES

Figure No. 42-11

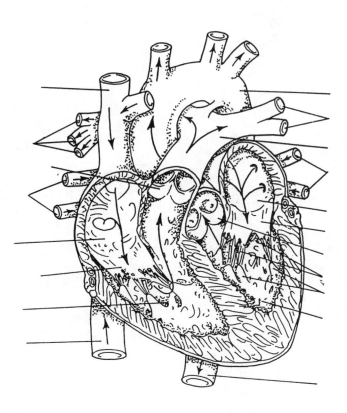

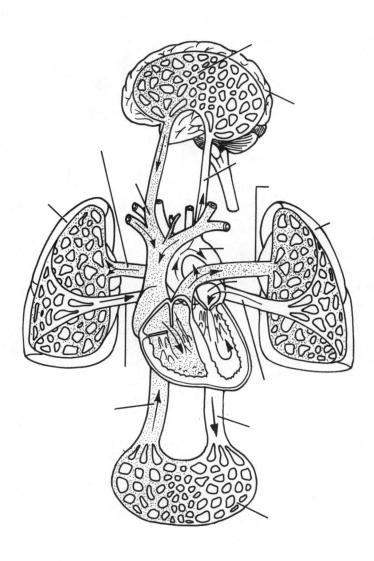

NOTES

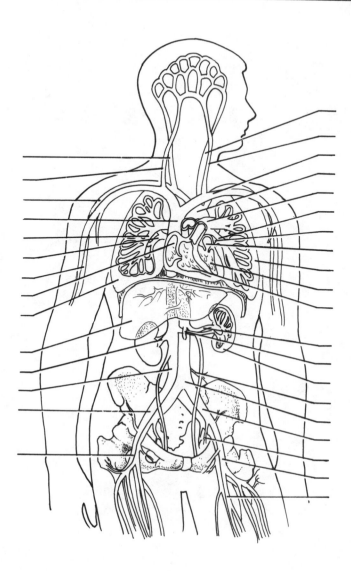

NOTES

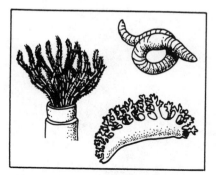
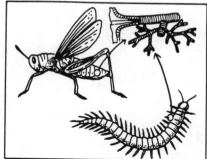
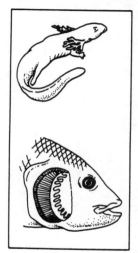
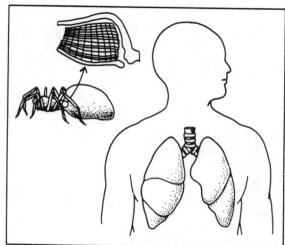

NOTES

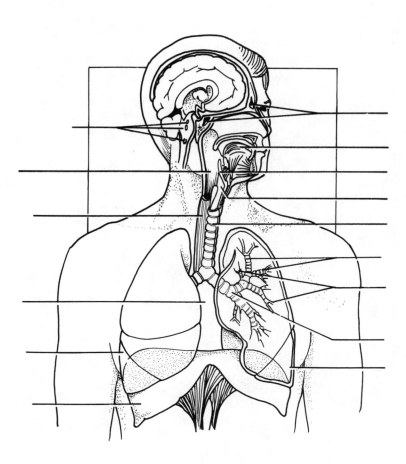

NOTES

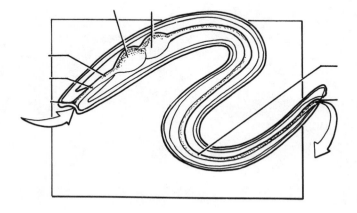

NOTES

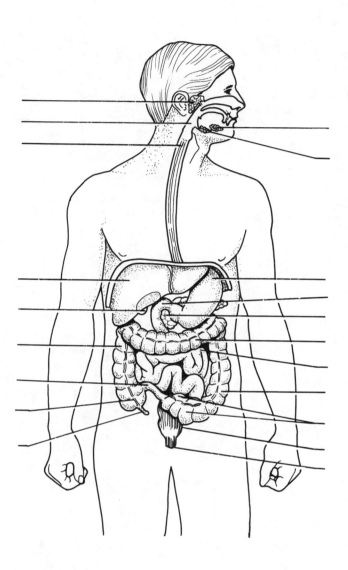

NOTES

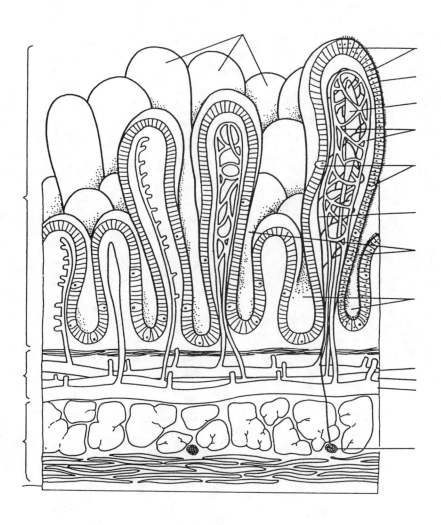

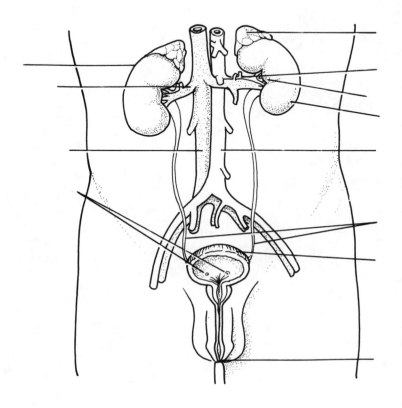

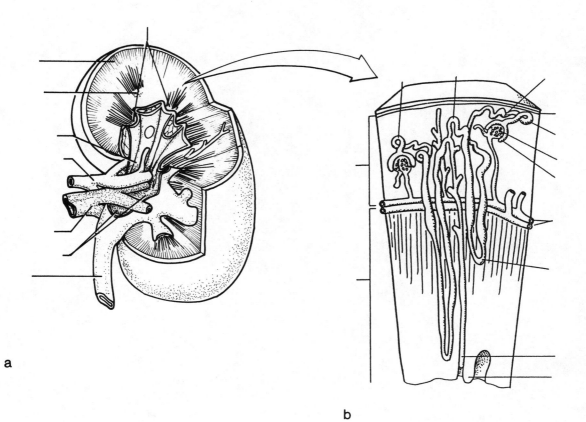

a

b

NOTES

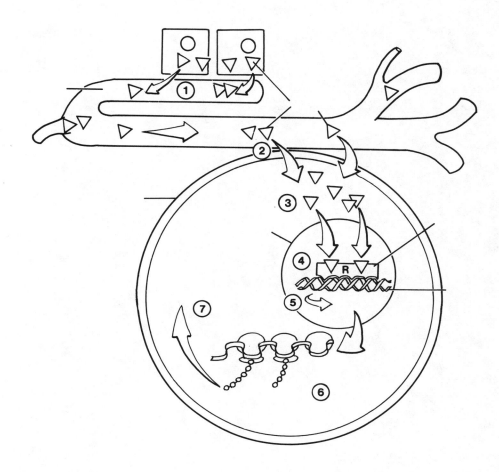

NOTES

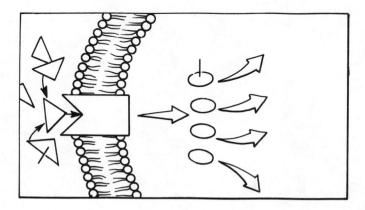

NOTES

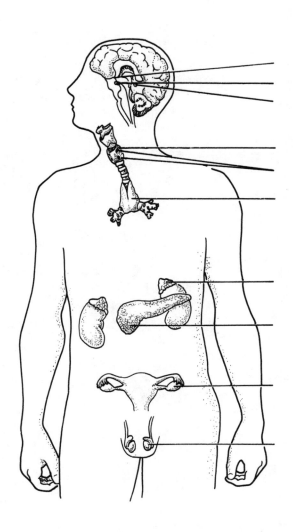

NOTES

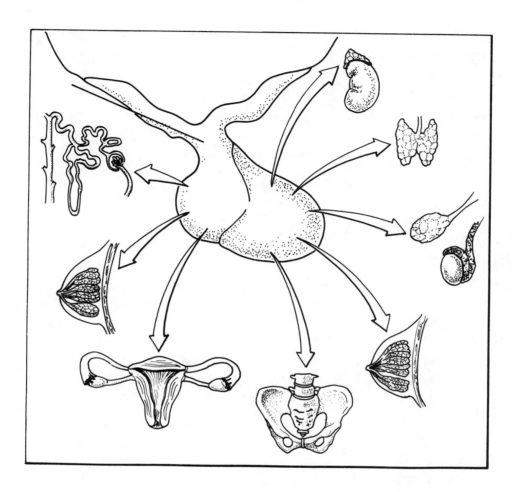

NOTES

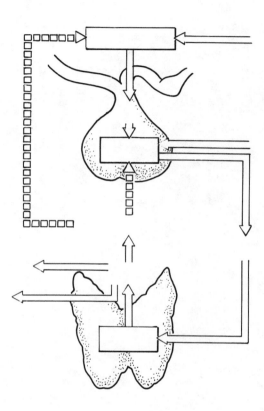

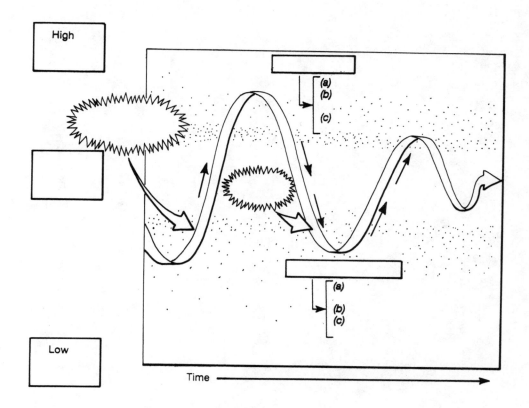

High

Low

Time

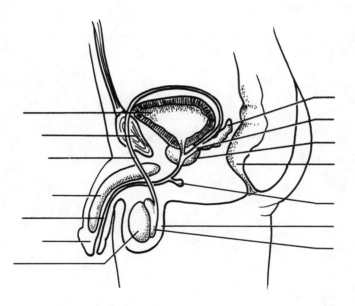

NOTES

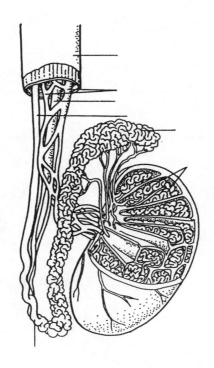

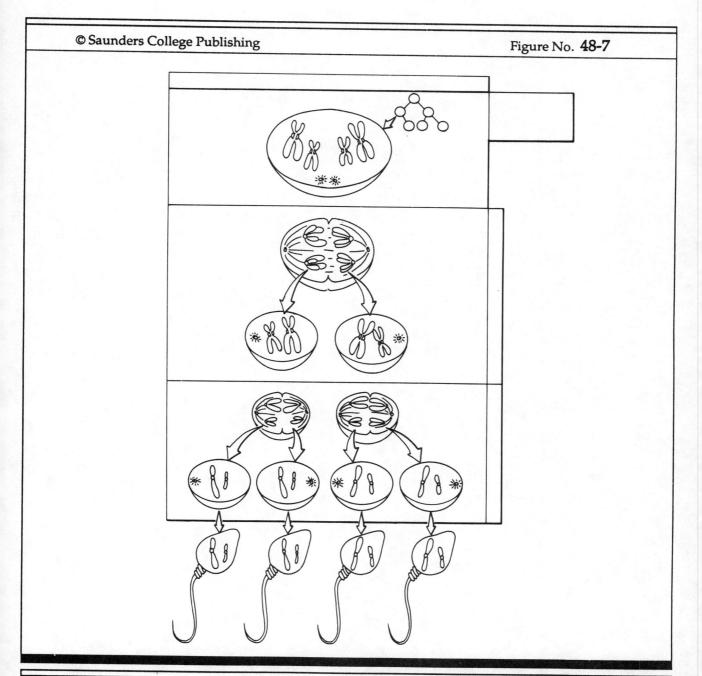

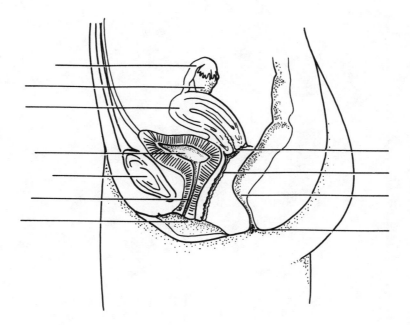

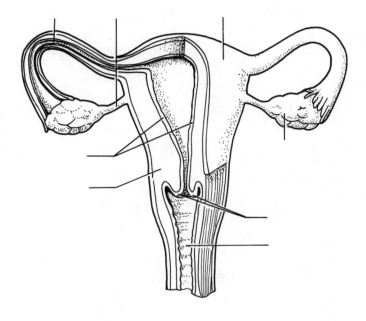

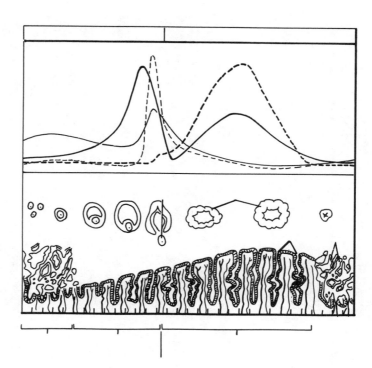

NOTES

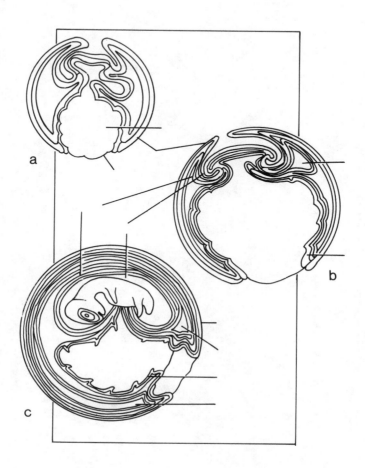

NOTES

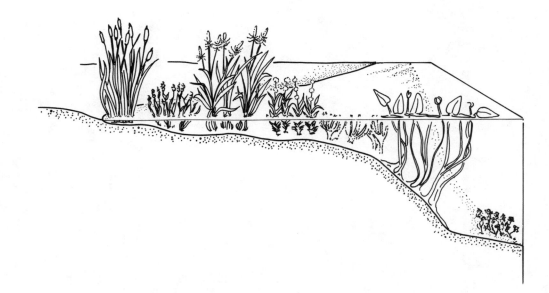

NOTES

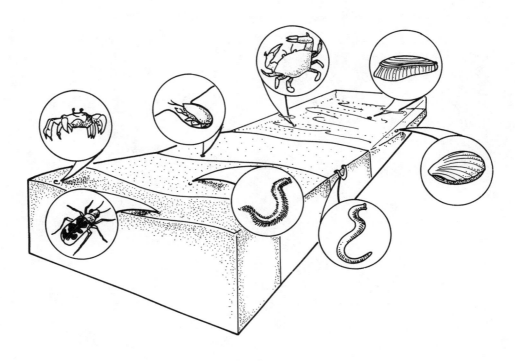

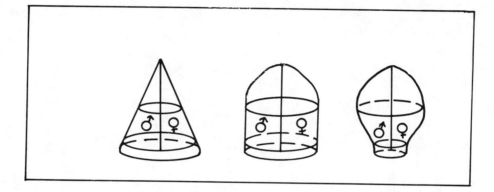

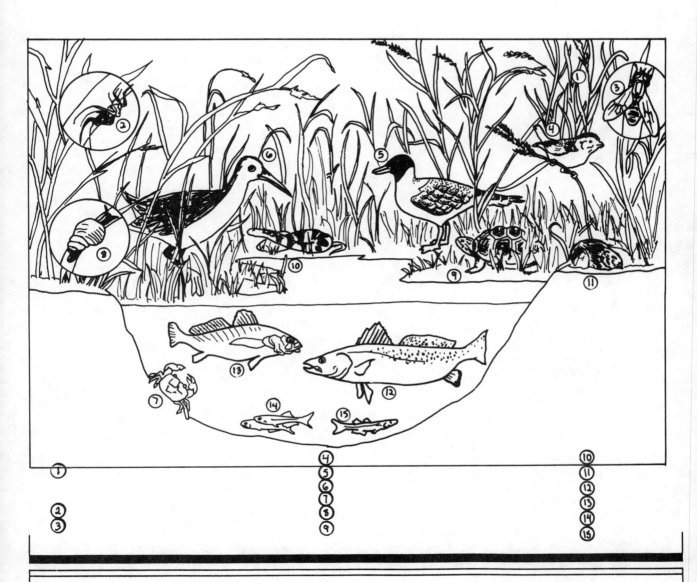

NOTES

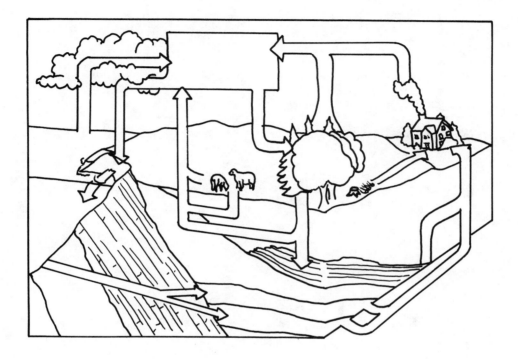

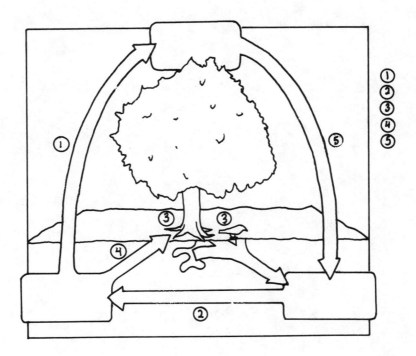

① ② ③ ④ ⑤

NOTES

NOTES

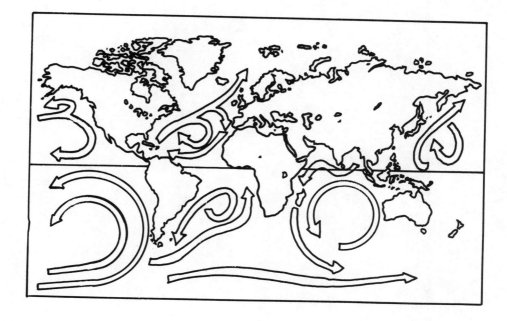

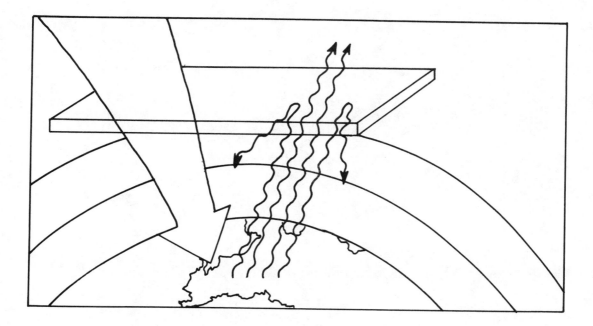

NOTES

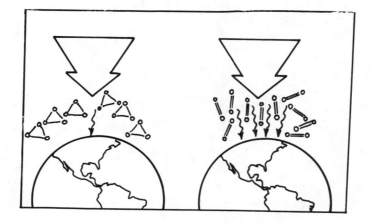

NOTES